Palgrave Studies in the Future of Humanity and its Successors

Series Editors
Calvin Mercer
East Carolina University
Greenville, USA

Steve Fuller
Department of Sociology
University of Warwick
Coventry, UK

Humanity is at a crossroads in its history, precariously poised between mastery and extinction. The fast-developing array of human enhancement therapies and technologies (e.g., genetic engineering, information technology, regenerative medicine, robotics, and nanotechnology) are increasingly impacting our lives and our future. The most ardent advocates believe that some of these developments could permit humans to take control of their own evolution and alter human nature and the human condition in fundamental ways, perhaps to an extent that we arrive at the "posthuman", the "successor" of humanity. This series brings together research from a variety of fields to consider the economic, ethical, legal, political, psychological, religious, social, and other implications of cutting-edge science and technology. The series as a whole does not advocate any particular position on these matters. Rather, it provides a forum for experts to wrestle with the far-reaching implications of the enhancement technologies of our day. The time is ripe for forwarding this conversation among academics, public policy experts, and the general public. For more information on Palgrave Studies in the Future of Humanity and its Successors, please contact Phil Getz, Editor, Religion & Philosophy: phil.getz@palgrave-usa.com.

Patrick Gamez

Posthumanism Meets Surveillance Capitalism

How to Delete the Manifest Image

Patrick Gamez
University of Notre Dame
Notre Dame, IN, USA

ISSN 2945-6592   ISSN 2945-6606 (electronic)
Palgrave Studies in the Future of Humanity and its Successors
ISBN 978-3-031-90769-2   ISBN 978-3-031-90770-8 (eBook)
https://doi.org/10.1007/978-3-031-90770-8

© The Editor(s) (if applicable) and The Author(s), under exclusive license to Springer Nature Switzerland AG 2025

This work is subject to copyright. All rights are solely and exclusively licensed by the Publisher, whether the whole or part of the material is concerned, specifically the rights of translation, reprinting, reuse of illustrations, recitation, broadcasting, reproduction on microfilms or in any other physical way, and transmission or information storage and retrieval, electronic adaptation, computer software, or by similar or dissimilar methodology now known or hereafter developed.
The use of general descriptive names, registered names, trademarks, service marks, etc. in this publication does not imply, even in the absence of a specific statement, that such names are exempt from the relevant protective laws and regulations and therefore free for general use.
The publisher, the authors and the editors are safe to assume that the advice and information in this book are believed to be true and accurate at the date of publication. Neither the publisher nor the authors or the editors give a warranty, expressed or implied, with respect to the material contained herein or for any errors or omissions that may have been made. The publisher remains neutral with regard to jurisdictional claims in published maps and institutional affiliations.

Cover illustration: © Evgeny Gromov / Alamy Stock Photo

This Palgrave Macmillan imprint is published by the registered company Springer Nature Switzerland AG.
The registered company address is: Gewerbestrasse 11, 6330 Cham, Switzerland

If disposing of this product, please recycle the paper.

ACKNOWLEDGMENTS

This work would never have begun without Joel White's generous and thoughtful invitation to submit something for a special issue of *Technophany: A Journal for Philosophy and Technology* on the concept of entropy. The germ of an idea quickly outgrew that format, and Joel was kind enough to release me from the obligation, so I'm doubly grateful to him.

The further development of this germ was helped along by thoughtful questions at the 2023 biannual meeting of the Society for Philosophy and Technology, as well as the History and Philosophy of Science colloquium here at Notre Dame (especially from Curtis Franks, Paddy Blanchette, David Batt, Anna Geltzer, and Ranjodh Singh Dhaliwal). Thanks are owed also to Michael Brown and Eugenia Torrance for the invitation.

The first version of this monograph fit my most natural mode of expression—a short, punchy, and not nearly well enough developed intervention. Encouragement and feedback from Shannon Mussett and Michael Ardoline, especially, was invaluable, as was thoughtful commentary from David Roden, Adam Rosenthal, Robert Goulding, and three anonymous reviewers from Palgrave, which led to the final version more than doubling in length. This of course means more than twice as many opportunities for me to be dead wrong. But their comments served solely to improve the book, and the errors are, as usual, solely mine.

I would be remiss not to acknowledge Dr. Mary Butterfield, simply because she asked me to. Curtis Forbes, too, but less enthusiastically.

This book has its origins in a particularly unhinged period of my life, which might have made a deep dive into the further edges of philosophical posthumanism even more appealing. At any rate, I definitely owe many thanks to Danielle Schweiss for her patience as I dragged it across the finish line.

Praise for *Posthumanism Meets Surveillance Capitalism*

"This book shows how both 'right' and 'left' accelerationism are ultimately undermined by contemporary machine learning-based AI. Gamez not only develops an original position in the field of posthumanism, but also shows how machine learning-based AI, through its confrontation with accelerationism, prometheanism and folk psychology, has direct and important philosophical implications."
—Dionysis Christias, *Academy of Athens, Greece*

"With impeccable scholarship and rare critical precision, Patrick Gamez takes the most important and radical arguments of recent Promethean and accelerationist thought as a foundation upon which he builds a compelling and timely critique of platform capitalism and of our responses toward it today and tomorrow."
—Richard Iveson, *Goldsmiths College, University of London, London, UK*

Contents

1 **Let There be Light** 1
   1.1 *Posthumanism Amidst the Ruins of Mind* 1
   1.2 *The Cybernetic Imaginary* 2
   1.3 *Daemonium Ex Machina* 7

2 **The Price of Fire** 11
   2.1 *Posthuman Orientation: On Myths and Gifts* 11
   2.2 *Delivering the Future* 17

3 **No Rx for NRx** 23
   3.1 *Land Speed Record* 23
   3.2 *Desiring-Production in the Circuits of Cybernetic Capital* 27
   3.3 *Death Driven: The Race for Intelligence* 33
   3.4 *Escape Velocity, or ...but We Were Promised Cyborgs!* 38

4 **Prometheanism and the Scientific Image of Man** 43
   4.1 *The Sellarsian Apocalypse* 43
   4.2 *Liberating the Space of Reasons* 51
   4.3 *Picturing the End of the Human* 63
   4.4 *Building the Robot at the End of Time* 74

5 How to Delete the Manifest Image: A Political Economy
   of the Mind   87
   5.1 *Technology's Ends and Odds*   87
   5.2 *The Machinic Revolution of Neoliberal Cybernetics*   94
   5.3 *Mont Pelerin Multivac: The Calculation Problem and the
        Stygian Intelligence of Platform Capitalism*   102
   5.4 *No Thyself: Black Boxes at the End of Theory*   113

6 Concluding Remarks: Defacing Thought, or,
  on Being Disoriented   131

References   139

Index   153

CHAPTER 1

Let There be Light

Abstract In this first chapter I introduce the basic aims of the text, before exploring the thermodynamic and informational themes of Asimov's "The Last Question." Accelerationism and Prometheanism have always been informed by the tropes of science fiction, and I argue here that the themes of reducing entropy and the transcendence of artificial intelligence, with their roots in the fledgling post-war science of cybernetics, inform both of these posthuman projects. I discuss the statistical mechanical interpretation of the Second Law of thermodynamics, and its connections to information theory, as well as Maxwell's Demon, taking some time to emphasize that classical Shannon information is a non-semantic measure of probability.

Keywords Asimov, Isaac • Thermodynamics • Information • Cybernetics • Science fiction • Sociotechnical imaginaries

1.1 Posthumanism Amidst the Ruins of Mind

This is a book about the end of the mind. In it, I engage with two intimately related forms of philosophical posthumanism, each of which is engaged in attempts to think about how certain ways of technologically extending, intensifying, or radicalizing aspects of the *mind*, whether desire or reason, might lead to the end of the *human*, which can then be seen as a mere limitation of these more interesting forces and capacities. As will be seen, I am deeply sympathetic to these views; there is something compelling, both aesthetically and theoretically, about the thought of an alien future that promises radical alterity. And this appeal isn't limited to the partisans of technological posthumanism, as Levinasian and deconstructive views of futurity, just for example, seem committed precisely to the arrival of such Otherness. Unfortunately, as will also be seen, I will be arguing that if one is sympathetic to this sort of posthumanism, the contemporary deployment of AI in platform capitalism doesn't merely herald the end of the human but also the mind that was to surpass it.

I make use of the tropes of science fiction, specifically those concerning the physical and informational dimensions of entropy and the thermodynamic arrow of time, and the way these are in turn incorporated into the cybernetic imagination of artificial intelligence, to test the promises of Prometheanism, a distinctive intellectual project at the intersection of posthumanism and transhumanism. Prometheanism is distinctive in both its commitment to rationalism and its commitment to a sort of technological inhumanism; rationality doesn't lie in human nature, or even in specifically *human* practices. Rather, reason is constructed by the dynamic development of technological systems, a future achievement. In this, Prometheanism is a mirror image of accelerationism, a controversial view on which the rapid development of our information technologies serves not to rationalize us but to intensify our desires past the point of being bound by, or intelligible to, the norms of reason.

I hope to show that by following the trajectory of these theoretical movements, what we find is neither the architecture of an artificial reason nor the explosion of that edifice by augmented desire. Rather, if we think of the mind, or mindedness, as a *technological* phenomenon, we have to take seriously the epistemic and normative dimensions of platform capitalism, wherein, I shall argue, the entire structure—the frame in which reason and desire find their proper place—collapses. I conclude by providing a brief gesture at the bleak sort of posthumanism that could emerge from the ruins of mind left in its wake.

1.2 The Cybernetic Imaginary

How should we orient ourselves toward the future? In 1956, Isaac Asimov published "The Last Question." Despite his long and stupendously successful subsequent career, it remained, in his words, "of all the stories [he had] written... [his] absolute favorite."[1] The short story is a fascinating reflection on what it would, or could, mean to orient oneself toward the future. What makes it such a distinctively contemporary reflection is that it is pitched thermodynamically, as it were. In it, humans have created something that we would now call "artificial intelligence," the Multivac. Not only would we call it artificial intelligence, but the Multivac resembles, in some respects, *today*'s version of artificial intelligence, that is, a machine learning algorithm that feeds on data, extracting its answers and correcting itself. It can answer any question put to it far more quickly and accurately than a human (ultimately, than any organic intelligence at all), and nothing else serves to correct it. On May 21, 2063, Multivac is asked the last question, not because it is the last question to be asked, but the last to be answered. The question, raised by an attendant playing hooky, is, in its essence, "Is it possible to ultimately reduce entropy?"

It is worth noting that Asimov has his character raise the question in the context of humanity's transition to entirely renewable solar energy; the massive computing power of Multivac allows humanity to escape, in advance, an Anthropocene disaster that doesn't even register in the early atomic age. Yet the orientation toward the future is an *energetic* one, a *thermodynamic* one, a thoroughly post-Newtonian one. The last question is prompted by the realization that even the sun is not *truly* renewable.

On the canonical, statistical mechanical understanding of the Second Law of Thermodynamics, entropy is a measure of disorder, of the tendency of systems to settle into a state of equilibrium, which, at the molecular level, ultimately means the tendency toward a state of completely random stillness. Unlike the forces of the Newtonian cosmos, which are essentially time-symmetrical and reversible, thermodynamic processes are essentially *asymmetrical* and time-directed: energy always dissipates, systems grow more disordered, and as long as we organized organisms and our posthuman descendants seek to maintain, or even grow, the vast workscapes of our civilization, the only fate that awaits is the heat death of the universe.[2]

---

[1] *The Complete Stories*, Introduction, e-book edition.
[2] I say "Newtonian" here to emphasize the wrench that thermodynamics throws in the mechanisms of the physical worldview in which it emerged. But both quantum mechanical and relativistic descriptions of the world yield time-symmetric fundamental laws as well. Contemporary physics does in fact have other hypotheses on offer for the end of everything, but the "Big Freeze" or heat-death hypothesis is the most widely held view in the field. See Émile P. Torres, *Human Extinction: A History of the Science and Ethics of Annihilation*, 403.

But if energetic, then also informatic. The same year that Asimov published his story saw the birth of the transistor, an electronic revolution that, in turn, would unlock the potential of the information age that had just begun to dawn, eight years earlier, with Claude Shannon's *Mathematical Theory of Communication*. In the latter, Shannon had not only ventured to quantify the content of communication, independent of anything like *meaning*, as *information*. I take some time in the following paragraphs to explain this notion of information in a little more detail. While for my purposes here, little hinges on the technical details, it is worth noting some important themes, and articulating something of the cybernetic imaginary within which Asimov and—as we shall see—the accelerationists and Prometheans work, and which may well constitute our shared horizon for conceptualizing thought.[3]

In its most basic sense, Shannon information is a representation of the relative probability of a state of affairs; information representing a highly unlikely state is greater than information representing a highly likely one. This probability is what is transmitted through a communication channel. Not for nothing, Shannon's information theory made central use of the concept of entropy. Indeed, it is often reported (though probably the anecdote is apocryphal) that von Neumann specifically told Shannon to appropriate the thermodynamic concept of entropy for information theory, precisely because the mathematical formulae governing the information conveyed by a channel and the entropy of a system are the same.[4] So, the universe of information has laws governing how much a message can be compressed, for example, or how much surprise a given channel of communication can produce, laws intimately related to those governing the slow settling of the cosmos into silence.

And yet, despite living in an "information" age, one in which unimaginably massive amounts of information are collected and constantly transmitted across vast distances, none of this information *means* anything. It is important to stress, here, that while colloquially (and historically) the term "information" has often denoted *specific semantic content*, as Shannon uses the term this is no longer the case. Indeed, insofar as one minimal

While the idea originates in the nineteenth century and thus has an air of "high modernist" about it, it nevertheless remains, for now, an inescapable horizon for thought.
    [3] I use the term "imaginary" here in the sense outlined by Sheila Jasanoff, as a set of future-oriented concepts and norms that shape intellectual, technological, and political projects. See "Future Imperfect: Science, Technology, and the Imagination of Modernity."
    [4] Lombardi, Holik, and Vanna, "What is Shannon Information?" 1988.

condition of semantic content, or meaning, is normativity, information *cannot* be meaningful; our everyday sense of misinformation or disinformation has no place in communication theory. Even the arguably most basic semantic norm—truth—falls away. A false message, even a necessarily false one, is just as informative as a true one. Rather, entropy, or the quantity of information conveyed through a channel, is the measure of *surprise*, or *unlikeliness resolved*, by the receipt of a message; if any message, independent of observer or origin, can be quantified, the only relevant measure is how *probable or improbable* the state it relates is.

This disconnect from *meaning* and, in turn, from *truth* has led philosophers to try to appropriate the communicative "flowing" of Shannon information while salvaging some normativity for it. The classic example of this is the epistemic conception of information put forward by Dretske, which ultimately makes information a *de re* veridical representation.[5] And the positive upshot of this is a conception of information—nowadays called "semantic information" —that can approximate the way we colloquially use the term in epistemic contexts. The cost of doing so, however, is that this is simply not Shannon information, the mathematical quantity that underlies the computing revolution of the twentieth century. This is made explicit, for example, in contemporary attempts to formulate a semantic notion of information, most notably that of Luciano Floridi, once referred to as "Google's Philosopher," which explicitly *require* a truth-constraint.[6] While Floridi is agnostic about the mechanism of this truth-constraint, Fred Dretske's strategy—and, to my mind, the most plausible one—is to stipulate a *causal* connection, from the information source to the recipient of the message (leaving aside the possibility that the source of information may not have any connection whatsoever to the content of information). However, as Richard Foley points out, this ultimately ends up cutting anything distinctive about *information* out of the account.[7] Foley's concerns are grounded in epistemic internalism and are beside the point here. Rather, what matters for us is that Shannon information is not described in causal terms.

---

[5] Richard Foley, "Dretske's 'Information-Theoretic' Account of Knowledge," 59.
[6] Cf. "Is Semantic Information Meaningful Data?" and "Semantic Information and the Correctness Theory of Truth." On the point about Google, see Robert Herrit, "Google's Philosopher."
[7] "Dretske's 'Information-Theoretic' Account of Knowledge," 165–166.

Why this short excursion into the concept of information and its epistemic foibles? To stress that *information as such* is not content. At the heart of it there is no meaning, no cause, only probabilities.

And the same non-causal, non-semantic probabilistic void lies at the heart of statistical mechanics, which purports to explain thermodynamics, in particular its perplexing Second Law, and, in turn, the arrow of time that points us unfailingly toward the lifeless stillness of absolute equilibrium. For how could one explain the irreversible, time-directed processes of thermodynamics solely in terms of the time-symmetric causal laws of mechanics, whether classical or quantum? Boltzmann's great insight was to shift our gaze from the macro-scale to the micro-scale, to the scale of molecules, the great collection of which was beyond our capacity to ascertain the position and velocity of each, which ultimately do not matter, but from which we could draw statistical, probabilistic inferences. The causal histories of each and every molecule, in total, are opaque to us. And that is fine because they are irrelevant; any could be reversed, in principle, moving from disorder to order, and yet entropy continues to increase. As mentioned, the link between the informatic and the energetic was no secret, as it could not be: the equations for entropy in each are identical, *save for the curious property* that their values were inverted: an increase in thermodynamic entropy corresponds to a decrease in information-theoretic entropy. As everything tends toward equilibrium, nothing becomes surprising.

Deciphering the connection between information and energy, probability and time, became central for a whole host of disciplines. In part this is because it lay at the heart of cybernetics. As a number of recent works have shown, the rise of cybernetics in the post-war period, its extension into disciplines from anthropology to biology to economics, and its eventual dissolution into a variety of separate sciences and concepts, like autopoiesis and cognitive science, have profoundly shaped our concepts and knowledge. In this sense, even if cybernetics is in some sense a "failed" science—the alchemy, or perhaps the Gnosticism, of the twentieth century—it still constitutes the vanishing horizon of our sociotechnical imaginary. In particular, the founders of cybernetics, John von Neumann, whose interaction with Shannon we've already noted, and Norbert Wiener, saw in the inverse correlation of information and entropy the possibility of creating a universal science of order and organization.[8] While the earliest

---

[8] See Ronald Kline, *The Cybernetics Moment or, Why We Call Our Age the Information Age*, esp. Ch. 1 and Ch. 3.

glimmers of cybernetics can be found in Wiener's wartime work on feedback loops and control engineering in anti-aircraft systems, the processes of self-regulation with which it is concerned mimic the local resistances to increasing entropy displayed by living organisms and it works precisely, as Schrödinger said of life, by feeding on "negentropy," or, in other words, *informing itself*. And so it was that "information" would become the basic "metaphor" of cybernetics.[9] The conceptual tensions between *meaning* and *information*, between the semantic and the causal and the statistical, between control and autonomy, also play out in the trajectory of accelerationist and Promethean posthumanism.

1.3 Daemonium Ex Machina

Insofar as the "Last Question" is in some ways a mirror of the "first question" of metaphysics—not "Why is there something rather than nothing?" but "How can we prevent what is from returning to nothing?"—we might expect it to carry with it some theological baggage. But, given the preceding, we should still not be surprised, at this point, that Asimov has his attendants ask the last question of a *computer*—if what becomes of Multivac can still be called a computer—rather, than, say, a traditional deity.

In the twentieth-century prehistory of artificial intelligence, Asimov is often taken to be a founding figure; his early stories about robots, of course, provide an image of a thinking machine, such that even if he was not a major technical contributor, he provided concrete inspiration—a possible future to which people like Turing and Minsky might contribute. And it's worth noting that, by the time he published "The Last Question" his thinking about thinking machines was already thoroughly cybernetic; the following year he wrote the preface for the English translation of Pierre de Latil's *La Pensée Artificielle*, a popularization of cybernetics, which Asimov describes as a revolution which will emancipate thought from slavery.[10]

So, on the one hand, Asimov's story simply exploits the ambiguous, cybernetic sense of entropy. The answer to the attendants' question is, in a sense, obviously affirmative (though not to the attendants who originally raised it): as (physical) entropy increases, (informational) entropy decreases. In the story, however, Multivac cannot answer the question

---

[9] *The Cybernetics Moment*, 56–61.
[10] "Foreword," viii.

immediately: it lacks sufficient data to give an answer. It is only after countless eons, as the universe settles into its final stillness and the machine has gathered not just sufficient but, it is implied, *all* data, that the answer is given. But on the other hand, and more importantly, the story transforms Maxwell's Demon into an electronic God. Maxwell's famous thought experiment was designed to explore the fact that, given its merely statistical grounding, nothing in principle prevents entropy being reduced. Given two chambers full of molecules in motion, and a frictionless panel between them, it seems possible that some sort of incredibly intelligent being could manipulate that panel, without expending any work or adding entropy to the system, such that fast-moving molecules would be permitted to pass to one side, and slower-moving ones to the other, thus decreasing the entropy of the system, and violating the Second Law of Thermodynamics. Asimov is here suggesting that Multivac might, in fact, be able to do something similar, and on a far grander scale.

We here have to ignore, of course, the subsequent development of thought about Maxwell's Demon. Like Descartes', it has invited attempts to refute the obviously false conclusion. It *has to be wrong!* It *must!* Leo Szilard would later hypothesize that any intelligent being would necessarily expend energy in order to *measure* the movements of the molecules involved, leading to a long and involved literature about the thermodynamic limits of physical computation, including the memory storage and the costs of deleting information. But what Asimov is exploring is the possibility of such a demonic being: an artificial intelligence who might come to know enough about the minutest details of things and thus, perhaps, as informational entropy reaches its lowest point, begin to reverse physical entropy.

And so, in the story, Multivac has gone through any number of metamorphoses and become "AC," a divine machine that has incorporated all humanity and subsequently transcended time and space—insofar as time is distinctively meaningful only insofar as it has a direction, and once the game of entropy has run down, it no longer does—and has an answer to the last question. The answer is affirmative, but it is not merely reported. Rather, the answer is a creative act, a divine performative, enunciated in the biblical declaration "Let there be light." Genesis takes place not—or not only—at the *beginning* of the cosmos. Artificial intelligence, in a sense, reaches *back* into time in order to bring the universe to its conclusion; while the increase of thermodynamic entropy gives us *our* limited sense of time-directedness, it is the transcendent work of the AI at the end of the world that sets this process in motion from its conclusion.

There are two central themes, drawn from "The Last Question," that will inform the subsequent chapters. The first concerns a general *orientation toward the future*. How should we confront the future, from where we are standing *now*, in a largely post-theistic context? That is, how can we make sense of the meaning of our lives, of anything, in a cosmos that, to the naturalist, appears to be running down to nothing, and in which no redemption is possible? The second revolves around the centrality of artificial, or inhuman, intelligence precisely in making sense of this future; however we ought to respond to the eventual heat death of the universe, doing so demands unlocking the mind from its meat casing and submitting to its alien wisdom. These two major themes are supplemented by what is undeniably a much smaller point in Asimov, but which will be relevant in the following: Multivac, Galactic AC, Cosmic AC, AC, the whole sequence of artificial intelligence that attempts to make sense of whether or not entropy can be decreased, is an alien mind not only because artificial but because connectionist, because it seems to be something like a deep learning algorithm, a post-finitude neural net. Asimov's vision of the divine mind is a cybernetic one, functioning by correlation and prediction, feedback, and response. The answer to the question is arrived at not by symbolic manipulation of meaningful content but by the process of correlating data, issuing not in a judgment but an act.

Science fiction shapes the sociotechnical imaginaries we inhabit, and the posthumanists we will encounter in the subsequent chapters continue to repeat these Asimovian, tropes. But that is just to say they—and, perhaps, we too —remain deeply enmeshed in the cybernetic image of the world. In the subsequent chapters, I focus on these themes of futurity, feedback, and AI in order to frame a discussion of an idiosyncratic, but nevertheless surprisingly influential, branch of posthuman thought, namely that of Prometheanism and accelerationism. In the next chapter, I will briefly introduce and contextualize Prometheanism, before moving on in Chap. 3 to its conceptual roots in accelerationism. Doing so will, I think, make clearer the continuities between these two branches, as well as the consequences of their shared commitments.

CHAPTER 2

The Price of Fire

Abstract In this chapter I introduce and contextualize Promethean posthumanism, pointing out its contrasts with transhumanism and resonances with other philosophical orientations, such as Bernard Stiegler's phenomenological pharmacology of technology. I identify the political or ethical heart of Prometheanism as a rejection of any "given" limits to the (post) human. Rather than seeing ethical or political limits as given, Prometheans are committed to the construction of rational technologies—indeed, of reason *as* a technology—that can provide us with normative guidance. I suggest that this may in fact lead to a profound normative disorientation (though I do not defend any philosophy of the given, and in later chapters will suggest that profound normative disorientation is our lot).

Keywords Prometheanism • Posthumanism • Dupuy, Jean-Pierre • Stiegler, Bernard • Brassier, Ray • Technological rationality

2.1 Posthuman Orientation: On Myths and Gifts

Posthumanists and transhumanists look to contemporary science and emerging technologies as means by which to transform, transcend, or discard the human condition in response to various exigencies. They attempt to orient us toward a future, grasped in terms of emerging technologies, existentially, ethically, and politically. So, for example, the transhumanist might think that, on broadly consequentialist grounds, the fact that we *can* hack the human body and its OS in order to radically extend human longevity or give (trans)humans control over the power and scope of their physical capacities means we ought to. The posthumanist, on the other hand, might think that our petty obsession with establishing human difference has led to the moral catastrophe of the Anthropocene, and make use of the growing body of knowledge in the life, earth, and information sciences to dislodge the human from its theoretical throne. In brief, the two can be distinguished in that, while transhumanists tend to exalt the rational core of the human in order to move beyond our current limits, the posthumanist "rescinds the privilege of the human" by "blurring… the boundaries between ourselves and inanimate objects (such as technological devices)."[1]

Prometheanism, I take it, splits the difference between these two orientations of thought. Like the transhumanist, the Promethean exalts rationality and intelligence as the key to a proper, or desirable, orientation toward the future. On the other hand, it absolutely rejects the priority of the human: reason and mind are by no means the exclusive property of the human, which may actually serve to impede them. For the moment we will focus on the former point.

The term "Promethean," of course, has a history. In the philosophy of technology, it's been used quite critically by Günther Anders to describe our troubling technological condition, and coined a second time by John Dryzek in the late 1990s to ambivalently describe a curious position in political-environmental discourse by, a combination of cautiousness—if not downright skepticism—about the severity of the impact of human activity on the environment, and optimism about the technological means at our disposal for increasing economic growth while managing or vitiating environmental externalities.[2] In this, they at least share in the spirit of the sort of Prometheanism I will be discussing here, which develops in the thought of Ray Brassier, Nick Srnicek, Alex Williams, Peter Wolfendale, Reza Negarestani, and Dionysis Christias. For these thinkers, Prometheanism aims explicitly at *remaking the world* according to rational

---

[1] Helen Hester, "SAPIENCE + CARE: Reason and Responsibility in Posthuman Politics," 68.
[2] *The Politics of the Earth: Environmental Discourses*, Ch. 3.

ethico-political ideals.³ And this remaking is *radical*. Not only is the external world to be remade, as is well under way in the built environment and the myriad infrastructural and energy systems that sustain it, but the inner world as well, that is, the self.

However, the apparently transhumanist commitment to remaking the world, technologically, in accord with reason is the expression of a deeper concern. Brassier opens his defense of Prometheanism by asking:

> What does it mean to orient oneself towards the future? Is the future worth investing in? In other words, what sort of investment can we collectively have towards the future, not just as individuals but as a species? This comes down to a very simple question: What shall we do with time? We know that time will do something with us, regardless of what we do or don't do. So should we try to do something with time, or even to time? This is also to ask what we should do about the future, and whether it can retain the pre-eminent status accorded to it in the project of modernity. Should we abandon the future? To abandon the future means to relinquish the intellectual project of Enlightenment.⁴

We will return to the meaning of Enlightenment in Sect. 3.3. But for the moment it is worth noting that Brassier echoes Asimov here; the Promethean motivation is a need to reckon with the future, to "do something with time," which, as we know, will utterly dissipate us. And the Promethean program is to revitalize Enlightenment through an inhuman rationalist, technological transformation, of both ourselves and our politics. We can orient ourselves toward the future by taking hold of time and remaking it, but better.

In Chap. 4 we will deal with more substantive Promethean commitments at length, but in this earlier work Brassier defends the Promethean attitude indirectly, by way of critiquing a critique of the radically transformative deployment of nano-, bio-, information, and computational technologies (the "NBIC convergence").⁵ He takes issue with Jean-Pierre Dupuy's Heideggerean-by-way-of-Arendt ontological worries with the

---

³Whether the "rationality" in question here is human or in some important sense "inhuman," that is, whether Prometheanism is simply a sort of humanism *manqué*, may be a point of contention. I take it that the Prometheans would think the latter; at the very least some of their most prominent representatives seem to be motivated strongly by the urge to move beyond humanism, as will be discussed in Chap. 4.

⁴"Prometheanism and Its Critics," 469.

⁵"Prometheanism and its Critics" p. 472.

development of these technologies. In brief, Dupuy worries that these are ways of obliterating or annihilating the delicate balance between *givenness* and *making* in the human condition.[6] One here sees the obvious resonances with Arendt, who interprets the Sputnik launch as emblematic of our shared drive to *escape the limitations* of the human condition, to replace or re-position ourselves in a different environment; to overcome the given equilibrium between the human being and its milieu.[7] But if the mere move beyond the biosphere within which life was given and made possible is not transgressive enough, it is worth remembering that Sputnik and the space race are the parents of the *cyborg*. Just three years after Sputnik, Manfred Clynes and Nathan Kline coined the term to describe the incorporation of exogenous machines into the human organism precisely in order to maintain homeostasis in the grand venture of space exploration.[8] Dupuy, being among many other things a historian of cybernetics, would be very aware of this.[9]

Dupuy's existential worry is ultimately that we will find ourselves without *given* limits. The opportunities provided by the NBIC convergence, just like those provided by earlier technologies, hinge on the fact that they make what was previously unavailable available; goals, processes, activities, intensities once barred to us become accessible. Fire made light accessible in the night; the Wright brothers, ultimately, made distant destinations available on demand. We still, of course, face roughly "given" limits in our lifespans and our cognitive capacities, however tweaked these might be by information technology and medical science and pharmaceuticals. The concern, rather, is with an ethos that sees any potential *limitation* as a *problem*, that is, for which there is no sense in which the given limits of our existence might in fact provide our lives with meaning, orientation, or guidance.

Speaking of Prometheus and pharmaceuticals, we cannot ignore here the work of Bernard Stiegler. We cannot address here all the nuances of the evolution of Stiegler's work, but his account of technology as a *pharmakon*, an ineradicable dimension of human existence that is both poison

---

[6] "Prometheanism and its Critics," p. 478. Brassier includes a very sensitive genealogy of the concept of human existence as finite transcendence, from Kant to Heidegger to Arendt to Dupuy; for this brief discussion I simplify the account but nothing crucial is lost from the points I wish to make.
[7] *The Human Condition*, pp. 1–6.
[8] "Cyborgs and Space," *Astronautics* 14(9): 26–27.
[9] Cf. *On the Origins of Cognitive Science: The Mechanization of the Mind*.

and therapy, presents as a more sophisticated version of Dupuy's concern. Stiegler contends that human being is constituted through an originary technicity, that is, there is nothing meaningfully human independent of technology, which is a sort of transcendental prosthesis for our being.[10] But this is consistent with there being meaningful differences between *empirical* technologies which expose different aspects of our organic life, and thus the different functions they perform, to augmentation and replacement, in turn transforming their relationship to ecological niches in which those organs and functions find a foothold. And we might find the loss of limitation in some arenas to be more troubling than in others, for example, in our access to baseline emotional responses to the world and others in it. Imagine callousness on demand.[11]

While Stiegler is not particularly concerned with the opposition between the *given* and the *made*—nor could he be, given what he takes to be the originary imbrication of being and technics—he nevertheless recognizes that our technological, or "organological," extensions of ourselves lead not only to the liberation and the expansion of our selves but also to "stupidity" or "automatism," in brief, to the same loss of meaningfulness and orientation in life that worry Dupuy.[12] Insofar as this pharmacological dimension of human life is ineradicable, we cannot hope to overcome it by merely submitting to the "given," which would never be purely given, but rather by taking a sort of "care," being careful not to fall victim to stupidity and attending to those aspects of our psychological, social, and institutional lives—namely, our instincts and drives—which are most important to us to conscientiously manage.[13] However, in constantly reiterating that the "organological," that is, the technological extension of our organic, functional engagement with the world, is necessarily distinct from its roots in "life," Stiegler nevertheless still normatively grounds this position in the organic, as the "extension of life by means other than life."[14]

The general point, for the Promethean, is that the organic or biological limitations we face are becoming increasingly *optional*. And, if this balance

---

[10] Stiegler presents the core of his views in the *Technics and Time* trilogy, the first volume of which contains an extended interpretation of the myth of Prometheus.

[11] See, e.g., R.S. Bakker "Crash Space" for a speculative investigation of the consequences of abolishing the limitations of our given moral-psychological responses to, e.g., the suffering of others.

[12] See, for example, *Nanjing Lectures*, 48–49; 56.

[13] *Nanjing Lectures*, 14–18.

[14] *Nanjing Lectures*, 9.

between givenness and making, between the enabling limits of our abilities and what we make of those abilities, is essential to us as *humans*, then our humanity is optional and the NBIC convergence is, thus, more opportunity than threat. For Brassier, as opposed to Dupuy, the cybernetic vision remains viable. We can and should open ourselves to a cybernetic future unbound by mere humanity.

Both Dupuy and Stiegler attempt to provide broad foundations for distinguishing those elements of our lived, technological milieu that should orient us, and provide meaning. For the former, it is the "given"; for the latter, it is some sense of "life."[15] But, at least with respect to Dupuy, Brassier notes that simply accepting the givenness of human finitude gives us no guide to how much or what kind of, say, suffering is simply to be respected as an ineliminable feature of that finitude and how much we might acceptably extinguish through technological means. Is cancer a given limit of human existence? Dementia? Death by exposure to the elements? What turns these limits to naked human existence into problems we might solve? Givenness or finitude actually provides no real criteria, no normative guidance at all.[16]

It is worth remembering here the price that Prometheus paid for bringing fire to humanity. The gift of fire, in the myth, is the gift of what we might now call technoscience, the tightly woven dual capacities for knowing and acting. For fire does not simply illuminate; just like life, it burns, heats, melts. It both reveals what is shrouded in darkness and in doing so—and *in order to do so*—consumes and transforms the environment in which it is set. And here we find another of Brassier's resonances with Asimov; his orientation toward the future, just like Multivac's, takes form in the context of confronting the inevitable heat death of the cosmos; it is intrinsically thermodynamic.[17] In his early opus, *Nihil Unbound*, Brassier precisely takes this extremity, the extinction of all thought, as the transcendental condition of Enlightenment reason, indeed, it seems, of all genuine

---

[15] Elsewhere I critique the very idea of an ethics founded on "life" as some sort of quasi-empirical object. See "Being Born Poorly: Steps Toward a Genuinely Postvital Posthumanism."
[16] "Prometheanism and Its Critics," 478–480.
[17] Joel White's remarkable article "How Does One Cosmotheoretically Respond to the Heat Death of the Universe?" has shaped my thinking about this matter, and allowed me to frame Brassier's thought in this way; any foolishness in what I draw from this starting point is of course due only to my own errors.

thought.[18] It unlocks mind from its temporal and corporeal limits. Moreover, it is this transcendental extinction that precisely demonstrates the *alienness* of thought, a foreign rationality that outstrips anything human but in which we might be able to partake. The end of a *meaningful* future allows for a *better* one. At this point, we should also recall that, in exchange for stolen power, Prometheus pays with his organs, chained to the rocks, set upon by Zeus' eagle day after day. In this sense, it is the *merely organic* that must be sacrificed, at least in principle, for the sake of a better technological future.

2.2 Delivering the Future

The Prometheans avow that normative guidance for the technological transcendence of the given can be provided by *reason*. So far, so Frankfurt School. But Promethean optimism differs from Teutonic pessimism in that it explicitly avows the radical *technological* transformation of both the world *and* we human beings in order to bring about these goals. At least for the early Frankfurt School, technology is governed by or an expression of a merely *instrumental* rationality, and, in turn, of a drive toward mere *efficiency* as opposed to a full-blooded sort of truly *practical* rationality that would be intimately linked to *organic life and its flourishing*.[19] In this, they have at least a slim resemblance to Stiegler (who, for his part, has very little to say about "reason").

For the Prometheans, to the contrary, *reason just is a technology*. That is to say, whether rationality is or is not an expression or outgrowth or necessary supplement or originary prosthesis to human organic life, it nevertheless might be radically hostile to it.[20] It seems worth taking a moment, here, to say a little bit about what it might mean to say that reason *just is* a technology. To begin, we should think of technology expansively, ignoring purported distinctions between technology in, say, Heidegger's sense, and some more restrictive and poetic sense of technē; it might be closer to Ellul's broad sense of "technics," including artifacts, systems, skills, rules, and techniques, but for the fact that Ellul gathers all these different

---

[18] I have dealt with *Nihil Unbound* at length in "Being Truly Wrong: Enlightened Nihilism or Unbound Naturalism?" and so will not engage in extended exegesis here.

[19] For what remains, to my mind, the most powerful articulation and defense of this view, see J.M. Bernstein, *Adorno: Disenchantment and Ethics*.

[20] To note that reason has, historically, come to be employed by *homo sapiens* is not to suggest that it is in the relevant sense an "outgrowth" of life. It has *autonomy* with respect to life.

elements together under the rubric of "the totality of methods rationally arrived at and having absolute efficiency... in every field of human activity."[21] The Promethean would deny that all technologies have the shared *telos* of efficiency, though they do seem to be committed to the view that information technologies are evolving toward increasingly powerful and increasingly alien forms of intelligence.[22]

Note as well that the claim is not merely that technology, or technological systems, is a necessary *condition* for reason or rationality.[23] In the same way that one can think of natural language as a technology—a technical system, embodied in phonemes and written signs, governed by normative rules for use—so one can think of reason. The obvious example of this view of language is, of course, the late Wittgenstein and his conception of language games, with its emphasis on the *pragmatics* of language, that is, on what our use of language *as a tool* does, but the idea of language as a technology, developed to meet specific communicative and practical goals, has more recently been taken seriously within linguistics.[24] Granted, these views are not normally adopted by *rationalists*, as the Prometheans all in some sense are, but that does not mean they are incompatible. Robert Brandom—from whom the Prometheans draw important lessons—at least implicitly places his rationalism in the Wittgensteinian-cum-pragmatist tradition when he adopts not only the metaphor of language "games" but of these games composing an urban landscape. However, contra Wittgenstein, for whom the city of language is dispersed and unplanned, a product of our natural history, for Brandom, rationality, the (skillful) game of giving and asking for reasons, is the *downtown* of this structure, from which the rest of discourse grows.[25] I am not here claiming

---

[21] *The Technological Society*, xxv.
[22] We will address this topic further in Sect. 5.1.
[23] In some contexts it is sensible to distinguish between reason and rationality; however, here I will be using them interchangeably.
[24] See, e.g., Daniel Dor, *The Instruction of Imagination: Language as a Social Communicative Technology*, Salikoko S. Mufwene, "Language as technology: Some questions that evolutionary linguistics should address," and Jan Koster, "Ceaseless, Unpredictable Creativity: Language as Technology." For a recent attempt at articulating—if not an identity—an intimate relationship between language and technology, see also Mark Coeckelbergh, *Using Words and Things: Language and Philosophy of Technology*.
[25] *Articulating Reasons: An Introduction to Inferentialism*, 14–15. Though I make reference here to Brandom, I take it that the Promethean view is consistent with those of Brandom's critics, for whom there are non-linguistic, non-discursive forms of intentionality, largely grounded in our pragmatic engagement with the world. On such views, insofar as our

that all Prometheans take mindedness to be a linguistic phenomenon, or to reduce mind or reason to language, but rather simply articulating a defense of the intelligibility of the idea of reason *as itself a technology*.

We shall see in more detail how all this plays out not only in Brassier, but in subsequent variations on Prometheanisms, such as offered by Peter Wolfendale and Reza Negarestani, in Chap. 4. For now, we can simply note that for Brassier, reason is simply a faculty—which is to say, a technique or capacity—"of generating and being bound by rules" and one subject to historicity for all that.[26] For Wolfendale, it is precisely the capacity to wield language, and technology more broadly, to abstract away from our limits and increase our practical and communicative capacities that is the heart of reason. In this, he remains consistent with Brassier; the "in principle generality" of rational cognition just is the ability to govern and bind ourselves to rules that transcend their particular applications.[27] Negarestani seems to go further than either: for him, rationality just is a self-correcting program for artificial general intelligence, working socially through language understood fundamentally as a technology.[28] We remain in Asimov's cybernetic imaginary.

But we might wonder if there is something circular, if not dissimulative, in exhorting oneself and others to radical technological transformation, with the proviso that this will be guided by reason, that this will be a continuation of the Enlightenment legacy, when this very reason is itself part of the ensemble of technologies and cannot remain untouched by that transformation. If reason is indeed a technology, then radical technological transformation is reason's radical self-transformation. On the one hand, there is something exhilarating about this prospect. On the other, however, we find ourselves precisely in the sort of situation that has long defied philosophical searches for rational criteria. The obvious analogue here is the now all-too-familiar Kuhnian paradigm shift; insofar as one scientific paradigm may have different norms of observation, forms of explanation, loss of prior knowledge, etc.,— that is, insofar as what counts as rationality has been transformed—it is not at all clear that there can be

---

pragmatic (and technological) engagement can constitute a fundamental sort of intentionality (and hence some form of mindedness), reason, as a "higher" form of intentional action or practice, may have even more of a claim to being "technological." See, e.g., Mark Okrent, "On Layer Cakes: Heidegger's Normative Pragmatism Revisited."

[26] "Prometheanism and Its Critics," pp. 485–486.
[27] "The Reformatting of Homo Sapiens" p. 62.
[28] See "The Labor of the Inhuman," as well as Ch. 2 of *Intelligence and Spirit*.

any sort of *rational* accounting for these shifts besides post-hoc *rationalization*. A similar analogue, in the ethical and existential register, might be the phenomenon of "transformative experience," which L.A. Paul has convincingly argued raises all sorts of problems for rational decision-making. So, for example, one cannot know what it's like to be a parent or—more fancifully—a vampire without becoming one. So one cannot compare being a vampire to being the self one currently is. But this is not *merely* an epistemic limitation.[29] This is so, at least in part, and at least in radical cases of self-transformation, because the values and interests that shape one's experience (and in terms of which one might make the relevant comparative evaluations) *will no longer be the same*. In a sense, one's practical reasons, upon transformation, will no longer be what they were; how, then, could one rationally choose between two lives, when they do not share a standard of rationality? In both cases, prospective guidance by reason seems foreclosed, even if one can retroactively rationalize one's choices. Despite his proud rationalism, in more recent work, Brassier himself seems to come close to acknowledging the issue of the opacity or indeterminacy of rational guidance for radical transformation when he claims that, for example, in assessing the dimensions of freedom and unfreedom we face, and which might be transformed into a better, freer future, we can only do so through "retrospective construction" of a history of disabling and enabling constraints.[30] As with Multivac, one cannot invent the future without looking backwards and inventing one's past, but nothing is *given*—not even by reason—that would provide guidance for how to do so correctly.

For now, at the risk of appearing theoretically conservative, especially in contrast to the Prometheans' bold affirmations of a dramatically different, and better, future, it is also worth remembering that Prometheus was punished with the repeated destruction of a very specific organ, namely, his liver. The price of fire is not simply the organic, but the capacity to filter out what is toxic from what is not.

For Stiegler, it is precisely this capacity to filter the toxic—to establish normative criteria—that allows one to project a meaningful future, as opposed to mere endless temporal becoming.[31] We don't need to agree

---

[29] We "must confront the existential dimension of our epistemic limitations" (L.A. Paul, *Transformative Experience*, 177). On becoming a vampire, see *Transformative Experience*, 1–5.
[30] "Strange Sameness: Hegel, Marx, and the Logic of Estrangement," 104.
[31] *Nanjing Lectures*, 109; 158.

with the details of Stiegler's account to acknowledge that, while the constant increase in entropy might provide time with direction, it is only in being able to conceive this temporal flux as *meaningfully oriented*, and something in which we can meaningfully intervene, that what is to come can *be a future*. This normativity provides time with an orientation and allows us—as Brassier says—to "do something with time." Prometheans hope to understand the normativity of time, and our future, in terms of a rationality in the process of constructing itself, which we will explain at more length in Chap. 4. In their attempts to deliver the future, I am going to suggest that what remains toxic at the heart of Prometheanism—what denies it a future—is its root in the perverse techno-libertarianism of Nick Land, that is, its accelerationist provenance.[32] However, I am not going to argue that Prometheanism is *wrong* because accelerationist, or—as folks have attempted to do with Heidegger and his heirs—that it is somehow "compromised" by the politics of its founders. Rather, I aim to show that the Promethean attempt to *sanitize* accelerationist posthumanism of its Landian past fails; they are twin cybernetic responses to the limits of humanism. In the following chapter, I provide an interpretation of Land's accelerationism, and of the central Promethean strategies for overcoming it; in Chap. 4, I engage with Prometheanism in more depth, sympathetic to its radical rationalism and inhuman extensions of Sellars' scientific naturalism, but with a view to pointing out its ambiguities and ambivalences. Critically, these ambiguities and ambivalences will only be resolved in Chap. 5, which confronts the AI imaginary of both accelerationism and Prometheanism with the exigencies of contemporary platform capitalism. Following the trajectory of this form of posthumanism to its limits allows us to understand how the entire shared framework in which they work is collapsing, both theoretically and historically. Our horizons might be expanded, for better or worse, by attempting to honestly gain our bearings in the ruins of the mind that this collapse leaves behind.

---

[32] To be clear, perversity here is no objection; it is part of what gives Land's early works their charm. But they are no less perverse for being charming.

CHAPTER 3

No Rx for NRx

Abstract In this chapter, I explain the work of Nick Land, the father of "accelerationism," a nihilistic and hypercapitalistic form of philosophical posthumanism developed through Land's readings of Jean-François Lyotard and Deleuze and Guattari in the 1990s. I interpret Land's thesis that capitalism is a primitive form of artificial intelligence "invading from the future" as fundamentally outsourcing our cognitive tasks in order to unleash *desire* from the limits of our all-too-human psychology and physiology. I then show how Land's turn to deeply reactionary politics in the twenty-first century is continuous with this project, and that Prometheans inherit from Land some central philosophical commitments.

Keywords Land, Nick • Accelerationism • Neoreaction • Artificial intelligence • Desiring-production • Race

3.1 Land Speed Record

Mark Fisher described his teacher, Nick Land, as being "our Nietzsche" but I think, rather, that Land is the closest thing to Heidegger we have in contemporary philosophy.[1] I don't mean to suggest something as ludi-

---

[1] "Terminator vs. Avatar," 341

crous as that the former has achieved anything near the level of influence and respect as the latter. And there is something perverse in the claim, insofar as Land, and the study of philosophy at Warwick more broadly in the last decades of the twentieth century, likely played no small role in the much less phenomenological and much more Deleuzian cast of much Continental philosophy in the UK. But bear with me. Heidegger's *Being and Time* was a genuine philosophical event, more or less fundamentally altering the trajectory of twentieth-century philosophy, drawing attention to underexplored aspects of existence and opening up the concreteness of life to ontological and hermeneutic interrogation. While it has long been known that he joined the Nazi party in 1933, this was not sufficient to render him *persona non grata* in the seminar room. It is only now, 50 years after his death, with the publication of the *Black Notebooks* that the *philosophical* depths of his Nazism have been revealed, perhaps forcing a reevaluation of those contributions.

One of those contributions was, of course, an ontological critique of technology as a mode of intelligibility. Heidegger, who wrote his most important works from his hut in the Black Forest, bemoaned that after the completion of metaphysics through technology, all that remains is "cybernetics," the universal science of feedback and control.[2] Land, on the other hand, spent the 1990s at the University of Warwick consuming methamphetamines and inventing a new cybernetic discourse, a genuine cyber-philosophy. To my mind, Land, along with Sadie Plant and others, embraced the emerging cyberpunk imaginary and incorporated it into philosophical thought in ways that remain unparalleled, attempting to think a radically transformed future while disinterring and rethinking the mid-century science of cybernetics. In this early work, Land attempted to rehabilitate the iconoclastic "philosophy of desire" that had briefly emerged in the French philosophy of the late 1960s and early 1970s as a rejection of orthodox Marxism and psychoanalysis: the libidinal economy of Lyotard and the schizoanalysis of Deleuze and Guattari. Benjamin Noys coined the term "accelerationism" for this strand of thinking; in the wake of 1968, taking *desire* to be the central concern of philosophical and political thought, it would end up being profoundly ambivalent toward

---

[2] *Four Seminars*, p. 63. See as well "The Task of Philosophy and the End of Thinking."

capitalism, which intensifies, unleashes, cultivates, and channels desire against any codification.[3]

Land's earlier writings were a rejection of the progressive/liberal normative structuring of both the subject and its society. His philosophical fathers, Deleuze and Guattari, had earlier attempted to overturn not only the Platonic idea of an isomorphism between self and city, such that imbalance or disease in one might rot the other, but also the more modern idea that society, or the state, might provide the realm of freedom in which the subject might realize or rationalize its nature, if only they were non-pathological. The latter idea, in some sense, underlies the history of both Marxism and psychoanalysis, the two main targets of criticism in *Capitalism and Schizophrenia*. In response, they would articulate a vision of self and society as part of a larger, wilder, ecology of machines, producing and working and interacting, channeling that work along various channels until those get smoothed over, reworked, and new paths torn through. Land's contribution—in some ways repeating Hayek, as we shall see in Sect. 3.2 and again in Chap. 5—consisted, partly, in radicalizing Deleuze and Guattari's thought by interpreting it in terms of *computing* machines, of information technology, that is, in terms of the networked, cybernetic world that was then coming into view, as if invading the present from the future. And, as we shall see, Land would explicitly claim that the future *was* invading the present; our very own capitalist Multivac reversing time to drag us into a future of desire so liberated that we would no longer need to desire at all.

So how is any of this like Heidegger? In three ways, the third of which is most significant. First, with Deleuze and Guattari as his Husserl and schizoanalysis as his phenomenology, Land would attempt to overcome the abstractness of his master's discourse, situating it in a concrete, historical moment, and attempting to move philosophy beyond its staid institutions and, if not into the anguished existence of *Dasein*, then into the dissolution of cyberspace:

> In the last half of the twentieth century, academics talked endlessly about the outside, but no-one went there. Land, by exemplary contrast, made

---

[3] Noys, *The Persistence of the Negative: A Critique of Contemporary Continental Theory*. See also Ch. 5 of David Hancock, *The Countercultural Logic of Neoliberalism*.

experiments in the unknown unavoidable for a philosophy caught in the abstractive howl of post-political cybernetics.[4]

Second, as we will discuss in Sect. 3.4 and Chap. 4, just as Heidegger shaped the itineraries of, e.g., Arendt, Gadamer, Jonas, Löwith, Marcuse, Agamben, and others, Land inspired a cohort of students who have, subsequently, transformed his insights while shaping important currents of thought. During his time at Warwick, Land supervised, taught, or otherwise influenced the generation of thinkers who would go on to articulate key ideas in contemporary cultural theory, philosophy, and beyond, including Mark Fisher, Ray Brassier, Iain Hamilton Grant (quoted above), and— more indirectly—Wolfendale and Negarestani. And, just as with Heidegger's students, part of their necessary philosophical task was to rehabilitate their master's basic insights as his work has, unfortunately, taken a turn.

Land would eventually leave Warwick in 1998, and the academy altogether, for Shanghai—the imagined "ethno-geographical core" of "Modernity 2.0," an aspirational capitalist-authoritarian regime—and, mostly, for Twitter.[5] In the decades since his exodus, he has become, for lack of a better term, a techno-reactionary.[6] His 2014 online essay "The Dark Enlightenment" trades in the neon-tinged affectations of his 1990s writing for ostensibly sober reflection, but the substance is not changed. What has changed is the explicitness with which Land now endorses the radically anti-democratic, libertarian, and—perhaps most surprisingly— racially charged consequences of his cybernetic schizoanalysis. We will lay out, all too briefly, this trajectory in the following subsection, addressing both how it remains well within the Asimovian imaginary and how it has shaped the itinerary of Promethean posthumanism.

---

[4] Iain Hamilton Grant, quoted in Robin Mackay, "Nick Land: An Experiment in Inhumanism." http://readthis.wtf/writing/nick-land-an-experiment-in-inhumanism/.
[5] "The Dark Enlightenment," Part 4e.
[6] Discourse around Land mostly takes place in cyberspace, on ephemeral websites and in a Theory blogosphere that has already become obsolete, but see Sandifer, *Neoreaction a Basilisk*; Hermansson et al., *The International Alt-Right: Fascism for the 21st Century?*, Ch. 7; and the references in note 19 above.

3.2 Desiring-Production in the Circuits of Cybernetic Capital

Almost every line of Land's essays scream silently for exegesis. For the sake of expediency I restrict myself here to this one passage, with a few supporting quotations. I hope the reader will be satisfied, though I acknowledge that a definitive-because-exhaustive reading is far beyond my scope here. In his essay "Cybergothic," Land lays out— fairly abruptly and not exactly transparently—three of his accelerationist commitments:

1. Anthropormorphic surplus-value is not analytically extricable from transhuman machineries.
2. Markets, desire and science fiction are all parts of the infrastructure.
3. Virtual Capital-Extinction is immanent to production. The short-term is already hacked by the long-term. The medium-term is reefed on schizophrenia. The long-term is cancelled.[7]

And we can get ourselves to a better understanding of Landian accelerationism, I think, by unpacking them.

As with Asimov and Brassier, the question of how to orient ourselves toward the future is central here. In Land's case, the answer is—unsurprisingly—a nihilistic acceleration. The future is *here*, exploding the present, and all we can do is hope to push through; "the long-term is cancelled" because there is no more future in the sense of some open, radically transformative horizon. As Thatcher would say, and Mark Fisher would lament, there is no alternative.

But why is this? Let's consider the first point, that "anthropomorphic surplus value is not analytically extricable from transhuman machineries." Land, following Deleuze and Guattari, is attempting to radicalize the concepts of desire and production in both a social and psychological register, moving beyond the normalizing restrictions placed on them by Marx and Freud. Where the latter thought there were normatively appropriate ways of structuring desire and production, in both the healthily integrated ego and the various relations of productions demanded by the development of the forces of production, Deleuze and Guattari see only proto-fascism, the cop inside our heads policing our hands and our genitals.

---

[7] "Cybergothic," 347.

Surprisingly, the most fruitful way to make sense of this is actually in terms of our commonsense folk psychology, with its basic distinctions between the representational (rational-cognitive) and motivational (desirous or conative) dimensions of psychic life. In the briefest terms, Deleuze, Guattari, and subsequently Land reject the idea, deriving from Plato, and transcendentalized by Kant, that reason, as a metonymy for our capacities for *representation*, should govern our *desires*, and in doing so transform the networks of conflicting and cooperating drives within us into stable subjects governed by the Reality Principle or historical relations of production.

Even on views, like the dominant Humean one, of psychic life on which desire is the ultimate driver of all human activity, reason still plays a role in governing and channeling those desires.[8] For Hume, what makes us who we are —what gives us our *character*—is the sedimentation of those desires by habit and their unification by instrumental reason. Kant radicalizes this, not only by giving reason its own motivating force through its own facticity, but by making the core of personhood the "I" that unifies our representations; not only can we govern our desires substantively, submitting them to rational ends, but that reason is the source of unity of those representations.

If Nietzsche attempted to explode this view of the subject by, rather, taking rationality, selfhood, and unity to be *effects* of the forces or drives that intersect us, it was, again, subsequently rationalized by Freud, for whom those forces can be integrated through the process of analysis (even if only ever incompletely, as a regulative idea). If the neurotic's drives represent the frustrated repetition of impersonal forces, analysis promises *wo es war, soll ich werden*; the "schizoid" who so fascinated Deleuze and Guattari, on the other hand, reveals the transience of the *ich*. In attempting to undo this Freudian domestication, they rather situate the subject, and its reason, not simply as a product of human forces but rather as part of broader, general ecology; there is nothing *human* about the drives that compel us, crossing our skin and bone and exiting us again. If, as they are constantly reminding us, Judge Schreber has a solar anus it is because those same cosmic forces traverse us. They are flows that do work in the animate and inanimate worlds alike, lacking any intentionality.

---

[8] For the Humean, reason cannot itself be motivating, but in its instrumental use it certainly can order and render most efficient the desires that one already has.

The key move here is their conception of desire as *desiring-production*. They no longer see the human being as fundamentally *lacking*, or the central expression of desire as *consumption*. That is to say, the completely impersonal, inhuman flow of energy that motivates us, moves us, and connects us to other people and things has no intention, no (proper) objects, and thus there is no way of organizing it that is not in some way an oppression or an accident. The flows of desiring-production, much like the flows of energy or information, make things do what they do: machines that work, in a non-normative and non-functionalist sense. What representation, what function, what normativity there is only emerges from the flows of desire that do work, produce value, digging furrows into social and psychological space, becoming coded, until a new set of desires deterritorialize them. On this view, the human itself is but an artifact, organisms merely transient vectors, of the productive force of desire.

But what have these "transhuman machineries" to do with surplus value? Why does the accelerationist crucially think the latter is "inextricable" from the former, as Land put it above? Here he has absorbed the libidinal economy of Lyotard on top of Deleuze and Guattari.[9] In the wake of 1968, Lyotard was disappointed by the intransigence of institutional Marxism in the face of pluralist calls for revolt and emancipation, of sexuality, decolonization, etc. The "libidinal turn" in all of their work shifts the locus of political attention from the *worker*, and indeed the human, to *desire*. Why hasn't the proletariat thrown off the yoke of capitalism in the nearly two centuries its inhumanity and countless indignities have been obvious? Because, for the libidinal thinkers, all those brutalities are also sources of *jouissance*, are productively invested by desire and hence a form of enjoyment. Why, on the other hand, has the second half of the twentieth century seen countless calls for the emancipation of desire, for the right to live as *pleases* one? So where for Marx, the "surplus-value" on which capitalism runs is a result of the exploitation of *labor*, which is the source of value, Land, following Lyotard, Deleuze, and Guattari, can only see this as the anthropomorphization, and needless limitation, of the flows of desire that attach and invest social and psychic space, creating "value" in the first place.[10]

---

[9] See Lyotard, *Libidinal Economy*, especially 108–114. It is worth noting, of course, that the text was translated into English by Land's student Iain Hamilton Grant.

[10] "For the fact of the matter, according to Lyotard, is that desire circulates endlessly around objects, surfaces, and bodies. In this connection, late capitalism is an immense desiring system" (Anthony Elliott, *Psychoanalytic Theory: An Introduction*).

The key commitments that Land would take from these libidinal thinkers are, of course, their profound antihumanism, construing desire as entirely continuous with the flow of energy at the planetary or cosmic scale, and, subsequently, their deep ambivalence—or occasional open celebration—of capitalism's destructive potential, exploding traditional hierarchies and proprieties and emancipating those flows of inhuman desire.[11]

And so, Land thinks, the infrastructure of these value-producing, transhuman machines involves "markets, desire, and science fiction." The key to bringing these ideas together is the interpretation of desire not only in terms of energy but of information. And to make sense of this we need to look, in a little more detail than Land ever did, at the neoliberal informatization of markets.

I follow Philip Mirowski here, in considering neoliberalism to be first and foremost an *epistemic project*, an enterprise of knowing, but a radical one, involving the transformation of what we know and how we know it.[12] This is important because, as we shall see, in some extended sense the first "Promethean" or "rational inhumanist" was none other than Friedrich Hayek. For Hayek, humans are radically opaque to themselves, and incredibly limited in their knowledge. This isn't *simply* a contingent feature of our individual histories, but rather a constitutive feature of the vastness and complexity of economic life: there *is no point* from which an individual could survey the totality of interests and their intensities that would allow them to coordinate their actions most efficiently. Rather, it is the *market* that serves as a *computer*, an information processor, constructing knowledge from dispersed and disparate inputs, and outputting prices that serve to coordinate the actions of agents and, in doing so, disclosing their desires and taking them as new inputs. In "The Use of Knowledge in Society," Hayek characterizes markets precisely as "mechanisms for conveying information."[13] Milton Friedman would later call the market not exactly an "analytical engine" but "an engine that analyzes" the world.[14] But, of course, for markets to work, they must be competitive, and their participants must be strictly rational. Concerning the latter point, as Hayek

---

[11] Later thinkers would, of course, criticize *la pensée '68* in its entirety for its proximity to neoliberal forms of management. See, e.g., Luc Boltanski and Eve Chiapello, *The New Spirit of Capitalism*, 97.

[12] "Postface: Defining Neoliberalism" & Mirowski & Nik-Khah, *The Knowledge We Have Lost in Information*.

[13] "The Ghosts of Hayek in Orthodox Microeconomics: Markets as Information Processors."

[14] Halpern, "The Future Will Not Be Calculated."

puts it in the opening passages, if we had all the relevant knowledge and (importantly) the *given preferences*, then understanding and predicting action, "the problem that remains," would be "purely one of logic." Thus, the designing and engineering of markets, the work of "formatting *homo sapiens*" as *homo oeconomicus*, serves as the realization of an inhuman rationality through the social institution of a machine intelligence.

What kind of knower is the market? An artificial intelligence. Which kind? Something like a machine learning algorithm. Following Mirowski's and Nik-Khah's work, Matteo Pasquinelli and, more recently, Orit Halpern have shown how Hayek was inspired by early accounts of neuroplasticity like those of Karl Lashley and, along with von Neumann, inspired Rosenblatt's early "Perceptron," embodying a cybernetic, connectionist view of cognition. More importantly, he construed the basic work of knowledge production as *pattern recognition*, and the algorithms of the market as something like a classifier algorithm, that works on unstructured data in order to provide distinctions and render the flux of human activity intelligible. So this neoliberal Prometheanism provides a vision of artificial intelligence that formats humans and renders them intelligible through the machine-learning market, with a view to disclosing the truth about their desires, in order to incite, cultivate, translate, and unleash them.

Here the science-fiction angle comes into play, connecting this infrastructure to the Asimovian theme of the entropy-eating AI of the future swallowing our present:

> Modernity discovers irreversible time - conceived as a progressive enlightenment tracking capital concentration - integrating it into nineteenth-century science as entropy production, and as its inverse (evolution).[15]

> Modernity marks itself out as hot culture, captured by a spiralling involvement with entropy deviations camouflaging an invasion from the future, launched back out of terminated security against everything that inhibits the meltdown process.[16]

> Machinic desire can seem a little inhuman, as it rips up political cultures, deletes traditions, dissolves subjectivities, and hacks through security apparatuses, tracking a soulless tropism to zero control. *This is because what appears to humanity as the history of capitalism is an invasion from the future by an artificial intelligent space that must assemble itself entirely from its enemy's resources.*[17]

---

[15] "Cybergothic," 315.
[16] "Meltdown," 445.
[17] "Machinic Desire," 338, emphasis mine.

We will explore all this in more detail in Sects. 5.2 and 5.3. But as neoliberalism piles up victory after victory, the future encroaches on us steadily. Nothing different will happen. Our destiny picks up speed.

The accelerationist can now see the increase in thermodynamic entropy that orients us in the present toward the future as the disguise worn by AI-Capital, which decreases informational entropy as our desires fuels it with data, inciting and encouraging them ever more, until they fully outstrip the temporary constraints of rationality and representation that make us "human." This is what Land, at the time, would call "Monopod" or "Human Security Systems," the political and mental institutions that would attempt to govern the flow of desire, preventing its ultimate entropic exhaustion and maintaining the semblance of humanity:

> Human history only makes it to Gibson's mid-twenty first century because Turing Security ices machine intelligence. Monopod anti-production inhibits meltdown (to the machinic phylum), boxing AI in synthetic thought control A(simov-) ROM, 'everything stops dead for a moment, everything freezes in place'. Under police protection the story carries on.[18]

For Land, in his radically indiscriminate Deleuzianism, the survival of the human itself is an artifact of fascist forces. If the desiring-machines were truly unrestricted, this would require a move beyond Hayek's market and their rationalizing coordinating power, an engagement with a technological platform which could incite the expansion of behavior without formatting its users as even instrumentally rational cognizers, to the point that we might become something else entirely. We will return to this idea in Sect. 5.4.

However, by 2014, Land no longer refers to the "Monopod" but rather, borrowing explicitly from the neoreactionary ideologue Curtis Yarvin, known better on Internet fora as Mencius Moldbug, to the "Cathedral," the dominant institutions of contemporary politics that aim to hold back the fragmentation of the political and, ultimately, the human.[19] While the register and tone of Land's writing have changed, I want to suggest that the motivations and endgame of right-wing accelerationism remain largely the same whether etched in glowing cybergothic neon or dull Gadsden yellow, namely, extirpating the human, a project of

---

[18] "Cybergothic," 347.
[19] "The Dark Enlightenment Part 1a" (https://www.thedarkenlightenment.com/the-dark-enlightenment-by-nick-land/)

self-liberation that is also a project of self-transformation so radical it might as well be self-annihilation. That is, I will be arguing for substantial continuity between Land's work at Warwick and his more recent, *prima facie* libertarian thought; I think this is not only accurate, but also helps make clearer how Prometheanism and accelerationism are mirror visions of the liberation of inhuman forces from their imprisonment within a humanity that stifles them.

3.3 DEATH DRIVEN: THE RACE FOR INTELLIGENCE

In an early essay on how the Freudian death-drive relates to Deleuze's desiring-production, Land interprets the death-drive not as a suicidal impulse to extinction but, with Freud and, perhaps surprisingly, against Deleuze and Guattari, as the "hydraulic tendency to the dissipation of intensities."[20] So, approaching their relation from the context of an emancipatory or critical psychoanalytic theory, if the "inhumanity" at the core of capitalism is a sort of machinism, or death, how can or should one resist it? For Land, one shouldn't. To elaborate, Land considers the question, broached first by Deleuze and Guattari in *A Thousand Plateaus*, of whether German National Socialism presents a historical example of the death-drive: society governed by a drive with nothing but nothingness as its object, dedicated to destroying itself rather than giving up that object. And Land's answer is: No. National Socialism represents, in a sense, the victory of the super-ego, a racial *ressentiment* that, precisely, blocks the flow of desire. And so he suggests we replace Deleuze and Guattari's question—"How do you make yourself a body-without-organs" —with the question "How do you make yourself a Nazi?"

> Above all, resent everything impetuous and irresponsible, insist upon unrelenting vigilance, crush sexuality under its reproductive function, rigidly enforce the domestication of women, distrust art, classicize cities to eliminate the disorder of uncontrolled flows, and persecute all minorities exhibiting a nomadic tendency.
>
> Trying not to be a Nazi approximates one to Nazism far more radically than any irresponsible impatience in destratification. Nazism might even be characterized as the pure politics of effort; the absolute dominion of the collective super-ego in its annihilating rigor. Nothing could be more politically disastrous than the launching of a moral case against Nazism: Nazism

---

[20] "Making It With Death: Remarks on Thanatos and Desiring-Production," 283.

is morality itself, heir to Europe's respectable history: that of witch-burnings, inquisitions, and pogroms. To want to be in the right is the common substratum of morality and genocidal reaction; the same desire for repression - organized in terms of the disapproving gaze of the father - that Anti-Oedipus analyzes with such power. Who could imagine Nazism without daddy? And who could imagine daddy being pre-figured in the energetic unconscious?[21]

Ever the edgelord, Land here basically makes the claim, so familiar to us from the post-Trump Right, that it is the Left, along with the liberal center, that are the real fascists. The death drive, the ultimate thermodynamic impulse to dissipation, which drives all of us, does not aim at genocide but at enjoyment. Nazism is in fact the perverse attempt to preserve the human order at all costs, but one that is instructive: if only they had learned to let go. National Socialism teaches us that, to escape the real forces of fascism, which repress our chaotic impulses, "revolution is not a duty, but surrender."[22]

Not much has changed. In "The Dark Enlightenment," Land continues with his Deleuzian crusade against the policing, or restriction, of desire, even if he has traded in the *soixante-huitard* rhetoric for the language of a right-wing blogger. Now, the key point, which he echoes from Peter Thiel, is simply that "democracy and liberty are no longer compatible."[23] And he is, by his lights, a champion of untrammeled liberty, where this is understood, precisely, as our surrender to our drives. And the so path to freedom is *exit*.

Drawing from Albert O. Hirschman's classic distinction between "exit" and "voice," and associating "voice" with both democracy and "representation," in its philosophical and political senses, Land thinks that the "Cathedral," a sort of broadly Left-liberal governing consensus, which aims to integrate political society through representation and redistribution—that is to say, which aims at giving "voice" through economic and social progressiveness—is ultimately oppressive. Moreover, in yet another echo of "Deleuze's anti-political economism" it is precisely "voice," discourse, and representation which are the tools of that repression, which is why those who believe in liberty must exit the political realm altogether.[24] And thus he approvingly describes Moldbug's "neocameralism," which

[21] "Making It With Death: Remarks on Thanatos and Desiring-Production," 284–285.
[22] "Making It With Death: Remarks on Thanatos and Desiring-Production," 287.
[23] "The Dark Enlightenment" Part 1a.
[24] "Making It With Death: Remarks on Thanatos and Desiring-Production," 264.

aims to replace political structures with economic ones, political sovereignty and power with economic sovereignty and power, replacing the nation-state with corporations. In the same way that Land's early work championed a hypercapitalist acceleration, for the later Land, the unfettered, information-processing market is once again a crucial instrument, and submission to it a necessary step, in the intensification and liberation of desire. And the exit in question here is exit from the human—an animal that, for land, can only exist in captivity.

The emancipation of desire through the escape from politics is not the only Landian theme that imbues his form of Neoreaction, or NRx, with its accelerationist flavor. It is also his obsession not only with artificial intelligence, in particular, but intelligence in general. In "The Dark Enlightenment," however, he views the inhuman ends of intelligence less through the cyberpunk sci-fi lens of the 1990s than from the perspective of cutting-edge biotechnological enhancement, from beyond the "bionic horizon."[25] He aims, as always, at an exit not only from politics but from the limits of human-all-too-human life, into a realm beyond: "think face tentacles."[26] The Lovecraftian reference here is pointed; Land is not interested in a eugenic biotechnology that might simply *augment* our human capacities in order to *optimize* our performance but, rather, in ultimately becoming something so transformed as to be unintelligible. There is, admittedly, something intoxicating about this idea. But the route he takes to get there is deeply troubling.

In some ways, the Dark Enlightenment essay is eerily prescient, e.g., explaining the otherwise confounding alliance of Silicon Valley libertarianism and Neo-Confederate white nationalism that played a role in bringing Donald Trump a presidency. Land's diagnosis is nevertheless unsettling. He spends much of the essay explaining how *race* and the conflict around it has made liberty impossible. Liberty, for Land, involves both emancipation and independence; since the Civil War bound together emancipation and integration, on the Union side, and independence and secession on the Confederate, there has been no space for freedom in the political realm. This unfreedom is enforced by the Cathedral's politically correct ruling ideology. Land is disturbingly sympathetic to the explicit racists, race-realists, and human-biological-diversity partisans who reject the imagined unity of the Cathedral in the name of racial difference, largely in

---

[25] "The Dark Enlightenment" Part 4f.
[26] "The Dark Enlightenment" Part 4f.

terms of purportedly natural variation in intelligence; independence here is the self-sufficiency of the (cognitively) superior race. And so the project of white nationalism dovetails with the libertarian dream, where the market will always reward the cleverest, if not the smartest, as part of the "crackerization" of the Right: the only option for both, in the face of the oppressive moral discourse and unbearable welfare of the Cathedral, is to *escape the political entirely*.

While Land was articulating his hypercapitalist posthumanism in the mid-1990s, cultural critics like David Golumbia, Richard Barbrook, and Andy Cameron were already pointing out the libertarian dimensions of the Internet.[27] Information technology had never seemed politically neutral, and Land was not the only one to notice its dynamics. But at the same time, libertarianism was undergoing a change. In "The Dark Enlightenment," with its explanation of the racialist intelligence hierarchy produced by economic competition unencumbered by government, and the need to secede from the state, Land is—perhaps unconsciously, perhaps not—simply catching up to Murray Rothbard, Hans-Hermann Hoppe, Peter Brimelow, and assorted other radically right-wing libertarians and anarcho-capitalists, committed to "cracking up" the democratic nation-state in favor of an "ethno-economy."[28] Like Hayek a student of Ludwig von Mises, by the early 1990s Rothbard was championing Pat Buchanan and David Duke while encouraging a right-wing populism to "repeal the twentieth century."[29] While that might seem backwards, at odds with a posthumanist ethos that aims at technological acceleration, the twentieth century to be repealed is the history of governmental impediments to economic liberty and, so the thought goes, technological innovation.

We will return to Rothbard and von Mises in Sect. 5.4. For the moment, we will simply note that, since the publication of "The Dark Enlightenment," with the rise of the Trumpian Republican Party, the odd coupling of a

---

[27] Cf. Golumbia, "Hypercapital," and Barbrook and Cameron, "The Californian Ideology."
[28] The work of Quinn Slobodian has been invaluable in bringing these threads together. See "Anti-'68ers and the Racist-Libertarian Alliance: How a Schism among Austrian School Neoliberals Helped Spawn the Alt Right," *Crack-Up Capitalism: Market Radicals and the Dream of a World without Democracy*, "The Ethno-economy: Peter Brimelow and the Capitalism of the Far Right," "The Unequal Mind: How Charles Murray and Neoliberal Think Tanks Revived IQ."
[29] See "A Strategy for the Right" and "Right-Wing Populism: A Strategy for Paleo Movement."

libertarian intelligence-obsessed Silicon Valley with a racialized, nativist populism has become the driving force in American right-wing politics. If accelerationism seemed somewhat fringe in 2016, the election of J.D. Vance to Vice President in 2024 brought these ideas directly to the White House, through the direct influence of Yarvin/Moldbug.[30] Indeed, Yarvin himself links the specifically Rothbardian inheritance of Austrian economics with renewed race science and "human biological diversity" as part of the neoreactionary rejection of Keynesianism *and* Chicago School neoliberalism and "human neurological uniformity."[31] This would be sufficient to worry, were it not for the rise of tech billionaire Elon Musk to a position of surprising (extragovernmental) influence; Land and Yarvin's dream of replacing political government with corporate government seems to be inching closer.[32] While not explicitly neoreactionary, Musk seems deeply committed to reworking *political* governance into *corporate* governance, and has long been a proponent of a number of views connected to eugenics, part of what Timnit Gebru and Emil Torres call the "TESCREAL bundle" of transhumanism, extropianism, singularitarianism, cosmism, rationalism, effective altruism, and longtermism.[33] In particular, his endorsement of longtermist values like the indefinite extension of the human race and overwhelming value of future generations, along with transhumanist commitments such as the production of artificial general intelligence, suggests something like a fetishization of intelligence, understood first and foremost as a genetic inheritance.

So we can see Land both foreshadowing a number of intellectual commitments of the contemporary New Right, and at the same time adjusting his thought to their actual historical trajectory. If, according to Land, the political proponents of these projects, like the Nazis in his earlier essay, fall short of truly submitting to their desires and being carried into the future by an inhuman mind, it is because their commitment to *humanity*, whether in misguided social conservatism and traditionalism or the technological enhancement of *human* intelligence, blinds them. Intelligence will not be enhanced through eugenics, which, of course can only conserve and purify

---

[30] James Pogue, "Inside the New Right, Where Peter Thiel is Placing His Biggest Bets"; Benjamin Wallace-Wells, "The Rise of the Thielists."
[31] See Moldbug, "ACG, KFM, and HNU."
[32] Thomas Beaumont, "Musk ascends as a political force beyond his wealth by tanking budget deal"; Michael D. Shear et al., "Elon Musk Flexes His Political Muscle as Government Shutdown Looms."
[33] Cf. "The TESCREAL Bundle."

an all-too-human genetic heritage. Rather, as we have seen, for Land, hyperintelligence will be realized outside of ourselves, through the AI-charged cybercapitalism that intelligently responds and caters to radically posthuman desire, aiming not at the extinction, as the Nazi might wish, of any particular race but rather of the *human* race.

In an interview with *The Guardian* on Land and accelerationism, Brassier expressed his disappointment with Land, claiming that "Nick Land has gone from arguing 'Politics is dead', 20 years ago, to this completely old-fashioned, standard reactionary stuff."[34] But this isn't correct. While the New Right is reactionary, it is naïve to think it old-fashioned or standard. But more importantly, on Land's view, the Last Question can only be answered, beyond politics and beyond humanity, by the *truly* Final Solution. Neither Peter Thiel nor Curtis Yarvin might recognize himself in this view, but it has always been Land's.

3.4 Escape Velocity, or ...but We Were Promised Cyborgs!

It is fair to say that Prometheanism, straddling the line between post- and transhumanism, is something like a "Left accelerationism." As we shall see, it maintains several of accelerationism's core commitments. But by the 2010s, many of Land's philosophical collaborators and disciples had grown wary of the direction of his runaway nihilism; even before his neoreactionary turn, it is likely that, in the wake of the 2008 financial crisis, at least, and the naked power of neoliberal austerity, the capitalist-AI invasion no longer looked as thrilling. They "explicitly sought to capture acceleration from... Land," attempting to reach escape velocity from an accelerating future gaining mass as it picks up speed.[35] This reclamation project has taken a number of forms, in the work of Mark Fisher, Nick Srnicek and Alex Williams, Ray Brassier, Reza Negarestani, Peter Wolfendale, and Dionysis Christias.[36] Though not precisely Promethean, Fisher's views are worth discussing (very) briefly in this subsection, in

---

[34] Andy Beckett, "Accelerationism: how a fringe philosophy predicted the world we live in," *The Guardian*, May 11, 2017 (https://www.theguardian.com/world/2017/may/11/accelerationism-how-a-fringe-philosophy-predicted-the-future-we-live-in).

[35] Hancock, *The Countercultural Logic of Neoliberalism*, 117.

[36] As far as I know, Christias is unique here in having no direct link to Land but nevertheless articulates a sophisticated and compelling Prometheanism.

order to provide some further context for the Prometheanism we are attempting to diagnose here.

Within the humanities, and certainly in Internet culture, Fisher is probably the most well known, and above all for his popularization of the concept of "hauntology." Though the term itself comes from Derrida, it has become Fisher's. In his work, the term comes to signify not just one but two ways of understanding the (temporal) present; insofar as, for Derrida, all presence is function of crucial absences, the temporal present is simply a relation to "what is *no longer* or *not yet*," in an attempt to

> think of hauntology as the *agency of the virtual*,... as that which acts without (physically) existing. The great thinkers of modernity, Freud as well as Marx, had discovered different modes of this spectral causality. The late capitalist world, governed by the abstractions of finance, is very clearly a world in which virtualities are effective, and perhaps the most ominous 'spectre of Marx' is capital itself...
> we can provisionally distinguish two directions in hauntology. The first refers to that which is (in actuality is) no longer, but which remains effective as a virtuality (the traumatic 'compulsion to repeat', a fatal pattern). The second sense of hauntology refers to that which (in actuality) has not yet happened, but which is already effective in the virtual (an attractor, an anticipation shaping current behaviour).[37]

Though Fisher never says as much, his use of "hauntology" is an immanent critique of Land, though bound to the same commitments. Where Land thought the future was invading the present, Fisher sees only the repetition of futures past, imagined futures that failed:

> The actual near future wasn't about Capital stripping off its latex mask and revealing the machinic death's head beneath; it was just the opposite: New Sincerity, Apple Computers advertised by kitschy-cutesy pop.[38]

The (ironically) vital, capitalist cyberpunk imaginary of the 1990s has been softened, replaced by nostalgia; we are hauntological insofar as we were promised cyborgs and instead received iPods and Roombas. For Fisher, this represents a *failure*: "Capitalism has abandoned the future because it can't deliver it."[39] And so Fisher wishes for a new, negative

---
[37] Fisher, *Ghosts of My Life: Writings on Depression, Hauntology, and Lost Futures*, 18–19.
[38] Fisher, "Terminator vs Avatar," 344.
[39] "Terminator vs Avatar," 346.

and—importantly—*affective* relation to the future or, rather, the frozen capitalist present; instead of the submission to desire that Land wished for, he hopes to reshape it through hate for the capitalist order that has disappointed us.[40]

Unfortunately, it's not clear that this is any objection to Land at all. Fisher, and others, may be upset by the "slow cancellation of the future," adapting the phrase from the resolutely anti-accelerationist Bifo, and the disappointment of our expectations. But from the accelerationist point of view, this is to be expected. Land said the same thing 30 years ago: "the long-term is cancelled," and the end of capitalism recedes from the horizon of political possibility. The rhetoric of revealing steel sinews holding together our obsolete meatsacks may be less appealing than it was in its moment (and still is, to many of us, today), but that doesn't mean the force of the substance of the view has changed. The future, for the accelerationist, is already here; it is the self-propelling nature of planetary technology that determines the shape of our present, but that doesn't mean that, from our point of view, it ought to change. All it does is capture our desire and our information, and all we do is dissipate.

Rather, what Fisher seems to want is to give a *different substantive end* to the accelerating process of technology; if AI is invading us from the future, perhaps it need not be *capitalist*. Perhaps we can dream differently, cultivating different ends, such that the acceleration of technology can be harnessed toward different ends.[41] The separation of technology and capital, which is to say, the divorce of desire and machine, is the aim. We will address this possibility at length in Chap. 5, but for now we simply note that it is unclear whether liberation is *compatible* with the dynamics of technological development or if Prometheanism ultimately must reject posthuman acceleration, reaching, like Benjamin, for the emergency brake.

Moving beyond Fisher's more affectively oriented thought, Alex Williams and Nick Srnicek explicitly position their thought as "Left accelerationism," picking up on Fisher's position, laying claim to a new and *better* future, one which is not here but rather needs to be brought into being:

[40] See also Gardiner, "Critique of Accelerationism," 34.
[41] Benjamin Noys, one of the most trenchant and persistent critics of accelerationism, also seems to think that resistance, or the redirection of the energies of contemporary technology, might begin a shift in our affective attitudes and stances; it is fundamentally aesthetic, a certain kind of political sensibility. See the conclusion of *Malign Velocities*, "People Are Afraid to Merge" (e-book).

The future needs to be constructed. It has been demolished by neoliberal capitalism and reduced to a cut-price promise of greater inequality, conflict, and chaos....What accelerationism pushes towards is a future that is more modern-an alternative modernity that neoliberalism is inherently unable to generate. The future must be cracked open once again, unfastening our horizons towards the universal possibilities of the Outside.[42]

Just like Land, they are searching for "exit," but this time not from politics, or a "Cathedral" that binds us to each other through restrictive norms, but from a form of capitalism that feeds on possibility and locks us into an eternal present. They explicitly name the goal that they aim for "a Promethean" politics of "collective self-mastery," which is to say, a kind of democracy.[43]

We can start to see here how both Land and his Promethean offspring are committed to the project of Enlightenment as a matter of *freedom*, even if the latter champion a form of democratic self-determination that Land explicitly rejects. This might seem surprising, for a number of reasons. After all, many take the legacy of Enlightenment to be a certain kind of *humanism*, a faith in human self-determination to ameliorate living conditions and in the capacity of human reason to liberate itself from illegitimate authorities that hinder its flourishing. But this is overly simplistic, and assumes that the great many goals of Enlightenment thinkers cohere into a single vision. But there is no such coherence. So, for example, Rousseau remains, for all his romanticism, an Enlightenment thinker for whom liberation lies precisely in the rejection of sophisticated rationality. Reason and emancipation pull apart. Likewise, Chantal Mouffe has demonstrated the conceptual independence—and, indeed, the deep *tensions*— between the Enlightenment ideals of liberalism and democracy.[44] So, it does not seem impossible that *humanism* and *rationality* and *emancipation* could all pull apart, while still remaining Enlightenment values. Land and the Prometheans, then, simply adopt different elements, under the broad Enlightenment aegis of "freedom." So Land's Dark Enlightenment is precisely that, an Enlightenment which ultimately rejects the human and any anthropocentric form of reason.

---

[42] "#Accelerate: Manifesto for an Accelerationist Politics," 362.
[43] "#Accelerate: Manifesto for an Accelerationist Politics," 360.
[44] See, e.g., "Radical Democracy or Liberal Democracy" and "Carl Schmitt and the Paradox of Liberal Democracy."

Indeed, both camps see Enlightenment as a project of *positive* freedom. Positive freedom is, in a sense, the *freedom to be who you really* are, a matter of self-mastery and self-realization. Often times, this is taken to imply that one must, indeed, be subject to important constraints; some of one's own impulses might prevent one from realizing one's genuine self. The Promethean *construction of reason*, as a collective epistemic and political project which we will explore in more detail in the subsequent section, can clearly be seen as a project of positive freedom. While Land's account of the unlimited liberation of desire, on the other hand, might look like a paradigmatic case of *negative* freedom, that is, the mere removal of obstacles to realizing those desires, this would be incorrect. Accelerationist self-destruction is a project of self-actualization—submission to the inhuman intelligence that is the true agent of history. The illusory pull of manifest desires is a smokescreen for the *drives* that integrate us into the circuits of machinic intelligence.

So, as I've noted, accelerationism and Prometheanism are mirror images of each other, responding to many of the same impulses and sharing a number of key commitments. The conflict between them is simply over what that self is. Human? Begrudgingly, contingently, at least for now. Mind? Perhaps—though, as we shall see in the next chapter, what that means is up for grabs. What sort of subject can technology build?[45] The answers to these questions also determine what is meant by the "Outside," and the "exit" that each searches for. Who will manage to escape, and from what?

---

[45] "Intertwined with this picture of liberated technological transformation is therefore the future of human beings. The pathway towards a postcapitalist society requires a shift away from the proletarianisation of humanity and towards a transformed and newly mutable subject. This subject cannot be determined in advance; it can only be elaborated in the unfolding of practical and conceptual ramifications. There is no 'true' essence to humanity that could be discovered beyond our enmeshments in technological, natural and social webs.... The postcapitalist subject would therefore not reveal an authentic self that had been obscured by capitalist social relations, but would instead unveil the space to create new modes of being. As Marx noted, 'all history is nothing but a continuous transformation of human nature', and the future of humanity cannot be determined abstractly in advance: it is first of all a practical matter, to be carried out in time. Nevertheless, some general notions might be entertained. For Marx, the primary principle of postcapitalism was the 'development of human powers which is an end in itself'. Indeed, the fundamental aim of his project was universal emancipation.... The immediate question is: What does this aim entail? The synthetic construction of freedom is the means by which human powers are to be developed" (*Inventing the Future: Postcapitalism and a World without Work*, 180–181).

CHAPTER 4

Prometheanism and the Scientific Image of Man

Abstract In this chapter I explore the distinctively Sellarsian cast of Promethean posthumanism, and how Sellars' naturalist rationalism provides thinkers like Brassier, Negarestani, Wolfendale, and Christias with a way of thinking beyond Land's accelerationism. I explain how they make use of Sellars' distinction between the "manifest" and "scientific images" to articulate a technological "space of reasons" that would allow for the expansion of the manifest image far beyond the bodily and cognitive limitations of the human. I go on to explore the tensions between reductionism and eliminativism about the mind that this view brings with it, before explaining how, like Land, the Prometheans seem committed to the construction of an alien, inhuman, artificial intelligence.

Keywords Prometheanism • Artificial intelligence • Sellars, Wilfrid • Functionalism • Eliminativism • Space of reasons • Artificial intelligence • Rationalism

4.1 The Sellarsian Apocalypse

We can now re-contextualize the extreme, inhuman rationalism of Prometheanism. On the one hand, it maintains the original accelerationist commitment to the "Outside," to the exit from or destitution of the human. Indeed, as we shall see, like Land, Prometheanism even thinks this Outside in terms of a submission to information technology, or artificial intelligence. What distinguishes Prometheanism from reactionary accelerationism—which is perhaps to say: from accelerationism as such—is its rejection of the centrality of *drives* and *desires*, that is, the essential rejection of either a psychoanalytic *or* schizoanalytic conception of intelligence. In Sect. 3.2, I framed Land's thinking in terms of the two basic components of folk psychology, namely, motivational and representational states. Of course, there seems something (precisely) *perverse* about viewing Land's cybergothic writings of the 1990s in terms of something as anodyne as "folk psychology," but the basic Landian hope is this: *desire*, when ripped from the rational and institutional channels that keep it restricted, can become *unintelligible*, whether on its own terms or due to the iterative, feedback-driven process of transformation that cybercapitalism might unleash. As transgressive as this vision is, it still has its root and motor in the (subpersonal) components of our psychology. But where Land wanted to reduce reason to a mere transient artifact of a desire that could outstrip human intelligibility, the Prometheans take the opposite tack, again producing a mirror image: the inhuman mind to which we submit is not driven by inciting our desires, consuming us as resource, but is rather the self-propelled, self-correcting engine of rational normativity. Against desire, *thought*.

The hope is, I take it, to maintain some basic accelerationist commitments—the primacy of the technological, for example, and the rescinding of the privilege of the human—without collapsing into fatalism about the capitalist capture of desire. The Prometheans do not simply fall into a facile representationalism, however, in contrast to a motivation-centric accelerationism. Rather, they adopt a distinctive rationalist position, one on which thought and its content are artifacts not of desire but of the normative commitments and sanctions that cognitive agents take on. That is, in order to make good on this rationalist hope, Prometheanism has taken a *Sellarsian* turn.

I won't here be *defending* this turn, though I am deeply sympathetic to it, as I am to the broader posthuman project of exploring the space of possibilities of thinking, being, and doing *beyond* the human. Unfortunately, as we shall find in Chap. 5, this sympathy only makes the consequences of contemporary AI for post-Landian posthumanism all the more unsettling. But, for now, the point is that if one is going to attempt to articulate a rationalist inhumanism, then one must articulate a vision of the mind such that it might be conceivably uncoupled from the human.

Now, on the one hand, there are multiple ways of construing the uncoupling of the mind from the human insofar as there are obviously a number of different views available to us of what rationality amounts to, and what a mind is, any and all of which are contentious for various reasons. One might, for instance, like Land, think that the uncoupling of the "mind" from the human is the becoming-unconstrained—and ultimately unintelligible—of desire, or intensification of a psychological state to the point of exploding the psychology to which it belonged. A different tack, but one I think largely in the same spirit, is taken by David Roden, whose speculative posthumans similarly stretch the concept of "agency" beyond rational intelligibility.[1] The point, here, is that Land and Roden take *rationality* to be the core of the human, and so uncoupling mind or agency from reason is the order of the day.

By contrast, for Brassier, Wolfendale, Negarestani, and Christias, the basic picture of rationality is largely derived from the work of Wilfrid Sellars and its development by later Sellarsians, such as in the inferentialism of Robert Brandom. The hope is that Sellars' work provides resources for developing an *inhuman* rationalism, on which reason is not in any important sense human and is capable of being divorced from the trappings of *homo sapiens*. To begin, it provides us with a view of the mind and its content that manages some degree of naturalistic respectability while at the same time managing to both separate the constitution of mental content from the biological heritage of the human and subordinate it to the practices of reason.[2] One can already see how this view would be attractive to one committed to rescinding the privilege of the human while advancing the rationalist legacy of the Enlightenment. However, as we shall see, especially in Sect. 4.3, Prometheans exploit an ambivalence in Sellars' work between functionalism and eliminativism with respect to the mind, an ambiguity that ultimately will be resolved in a far bleaker vision in Sect. 5.4.

---

[1] See, e.g., Roden, "On Reason and Spectral Machines: Brandom and Bounded Posthumanism."

[2] This is not to impute a magical, nonbiological *genesis* to the game of giving and asking for reasons, which surely emerges, as Wittgenstein puts it, from our "natural history" (*Philosophical Investigations* 16^e/§25). Rather, it is to point out its conceptual, and existential, *independence* from its inherence in our (for-now) animal life.

In the rest of Sect. 4.1, I will present the core Sellarsian contrast between the "manifest" and "scientific" images. In Sect. 4.2 I explain how the Prometheans make use of Sellars' concepts of the "myth of the given" and the "space of reasons." This in turn will help us grasp how the Prometheans view the emancipatory potential of Sellarsianism: if the practices of reason are separable from human embodiment, then we can envision not only the expansion and transformation of those practices but also the systems that perform them. However, this liberation presupposes a sort of functionalism about the manifest image of the mind which is at odds with some elements of both Prometheanism and Sellars' thought, which we will discuss in Sect. 4.3, before moving on in Sect. 4.4 to the Promethean prospects of artificial intelligence for the emancipation of the space of reasons, within the cybernetic imaginary.

Sellars is famous, of course, for distinguishing between the "manifest" and "scientific" images of human being. At its core, the language of the two "images" is but one more way of articulating the tension, so powerful in modern philosophy, between two seemingly incompatible descriptions of the world, each of which seems to have an authoritative claim over us: an intuitive philosophical humanism and an austere, nihilistic ontological naturalism. In other words, the problem is the by-now familiar one of reconciling the apparent force of the normative bonds in which we find ourselves with the stripped-down, norm-free vision of the physical world provided to us by the natural sciences.

The problem of finding *values* or *norms* in an indifferent cosmos is not the same thing as worrying about the place of seemingly super- or non-natural entities in a natural world. So, for example, one might think that there are immaterial minds, capable of intentional thought, propositional attitudes, or what-have-you, and still think the natural world devoid of norms and values. Dualism and nihilism are perfectly compatible. But, as we shall see, on the Sellarsian view, the two issues are intimately related. The mind is shot through with normativity, and just as we shall not find norms or values catalogued in the inventory of nature—that is, in the "scientific image" of the world —we will not find minds either.

Here I admit that my interpretation of Sellars is, perhaps, a "violent" one, and is shaped by the Sellarsian*ism* I want to articulate, the Promethean version of Sellars. But this is to be expected; not only the details of Sellars' work, but his aims, the positions to which he is responding, etc., remain matters of debate, if not outright controversy, in the literature. While his legacy in philosophy has been shaped by "right wing" and "left wing"

Sellarsians, the former committed to Sellars' radical naturalism in which cherished metaphysical notions like mind and meaning are replaced by complex physical systems, and the latter rather to a constitutive, and thus irreducible, holistic vision of rationality and mind, I present a Sellarsianism committed to both.

As Bas van Fraassen puts it, Sellars' project is eliminativist, an "apocalyptic vision."[3] But that is not the whole story. It is equally a conciliatory vision. His question is "to what extent does manifest man survive in the synoptic view which does equal justice to the scientific image which now confronts us?"[4] How does one reconcile oneself to an existential, semantic apocalypse?[5] And what question could better express the theoretical project of rationalist, posthumanist Prometheanism?

The "manifest" image seems to denote something like the "lifeworld" discussed by Husserl and Habermas, the world as it manifests or shows itself to us effortlessly.[6] For now, the basic intuitive sense of a description of the everyday world of experience, the world in which it is correct to say that one *thought, correctly,* that the sun rose in the morning, will suffice. However, it's important to note that the manifest image is, at least in part, *an image*, that is, a *description*, or the *framework* for a certain sort of description; it does not refer directly to "experience" or the "objects" or "world" of experience. Nor could it. As we shall see, given Sellars' commitment to scientific realism and his rejection of explanatory appeals to experience, some direct connection to an intelligible reality is foreclosed to him. Moreover, the very idea that there is a *tension* or *competition* between the manifest image and the scientific image, as Sellars thinks there is, where the scientific image is a *theory-based* framework, implies that the manifest image is at least in part in the same business of providing explanations and populating ontologies. For now, however, I will just note that Sellars' main interest in the manifest image lies in its rationalist-cum-existentialist conception of *persons*. This is, as he puts it:

---

[3] "Wilfrid Sellars' apocalyptic vision."
[4] "Philosophy and the Scientific Image of Man," 15.
[5] The term "semantic apocalypse," meaning (ironically) the end of meaning-talk, is owed to R.S. Bakker.
[6] This similarity is discussed at length in Dionysis Christias, *Normativity, Lifeworld, and Science in Sellars' Synoptic Vision*, esp. 29–33, though for the most part Industrials are, at best, indifferent to phenomenology.

the framework in terms of which man came to be aware of himself as man-in-the-world. It is the framework in terms of which, to use an existentialist turn of phrase, man first encountered himself-which is, of course, when he came to be man. For it is no merely incidental feature of man that he has a conception of himself as man-in-the-world, just as it is obvious, on reflection, that 'if man had a radically different conception of himself he would be a radically different kind of man.[7]

For Sellars, to be human is to have an image of oneself *as* human. But we can, as good Sellarsian posthumanists, free ourselves of the "human" here. Rather, we can simply say that to be *a person* is to have a sense of oneself as a person, to *structure* our comportment in the world as comportment toward persons.

As mentioned, the existentialist flavor of the manifest image is tempered by Sellars' rationalism. The framework of persons is the framework in which we find others, with whom we can engage, whose behavior is governed by *rules*. As he puts it in an earlier essay, describing the manifest image in different terms:

> a structure of rule-regulated symbol activity, which as such is free, constitutes a man's understanding of this world, the world in which he lives, its history and future, the laws according to which it operates, by meshing in with his tied behavior, his learned habits of response to his environment. To say that man is a rational animal, is to say that man is a creature not of habits, but of rules. When God created Adam, he whispered in his ear, 'In all contexts of action you will recognize rules, if only the rule to grope for rules to recognize. When you cease to recognize rules, you will walk on four feet.'[8]

To be a *person* is to *be governed by rules*, whether this means that one consciously *follows* those rules or, rather, is simply properly *evaluated* in terms of them, perhaps *as if* they were following rules. And it is important to remember that *rules* are not the *laws* that govern the natural world, that constitute the ironclad, necessary relations of the scientific image. They are *normative*, and essentially *rationally* so. So, if the manifest image is the framework in which we encounter each other as persons, it is the framework in which we encounter each other as *accountable to each other in terms of the rules of rationality*.

---

[7] "Philosophy and the Scientific Image of Man," 6.
[8] "Language, Rules, and Behavior," 295.

This framework is, as Dionysis Christias stresses, a *categorial* framework.[9] That is to say, the dimension of the manifest image in which we find persons is not an *empirical* one (though, as we shall see, the manifest image still has empirical dimensions). Persons, as subjects of reason, are also a distinct type of *logical* subject. In elaborating on this idea, Sellars entertains the idea that the manifest image is the refinement of an "original image" in which *all* things confront us as persons, not because we recognize certain properties by which they might be identified as such, but because personhood provides the categorial structure of this image.

The "scientific image," by contrast, is shorthand for the human being, and the world it inhabits, as described by the languages of the natural sciences, from which the mental and the normative are, perhaps troublingly, absent. It is much more straightforward to characterize. The scientific image comprises not just categories like objects and events, but the *actual* kinds of objects delivered to us by the natural sciences. It is an image of subatomic particles and weak fields, but not a world of minds, in the sense of bearers of propositional attitudes or contents. It is in this sense that Sellars is a naturalist; the scientific image is a naturalistic ontology, populated only by naturalistically respectable entities, which the mind, in the classical sense of a thing that *thinks*, is perhaps not.

But Sellars' use of the language of the manifest and scientific images is valuable for his articulation of the conflict between the two, a conflict that arises only when *both* are considered scientific images, that is, as competitors in the business of describing the world. If we consider the manifest image as an *empirical* refinement of the original image, then we *can* view both images as involving *theories* that are meant to *explain* the behaviors of things. And this tracks Sellars' own admission that one can view the manifest image not *only* as a categorial refinement of the manifest image, but as an empirical one.

So Sellars does indeed claim that, besides the categorial, there are empirical, explanatory dimensions of the manifest image. It has its own forms of description and explanation, and is answerable to the facts about how things are. Crucial, however, in distinguishing the manifest from the scientific image are the different forms of explanation belonging to each. While explanations in the scientific can make use of models with hidden variables—so-called "theoretical posits," objects and causes that may not be accessible to us—the manifest image makes use of "correlational

[9] *Normativity, Lifeworld, and Science in Sellars' Synoptic Vision*, 28–33.

induction" only, establishing regularities—perhaps even nomic ones—between antecedent conditions and their apparent consequences.[10] So the manifest image, with its rationally governed minds, is built up from originary concepts of *persons*, as unified rational (human) entities, and *thoughts*, the latter being concepts-by-analogy with verbal speech.[11] These internal thought episodes can explain behavior, being linked to *actions* by *inferences and motives*, arrived at by correlation and induction.

With the rise of empirical sciences, however, especially the biological and cognitive sciences, this explanatory, empirical dimension of the manifest image comes under threat; the physical, causal explanations of natural science threaten to crowd out the posits of thought. The latter find no place in a disenchanted world, their work given over to neurophysiology and biochemistry. And, if that work is so given over, Sellars takes it that any sort of *ontology* corresponding to the explanations of the manifest image, that is, any *theoretical posits* that might be made on the basis of our understanding of the manifest image, must similarly give way. Sellars is committed to scientific realism, a sort of ontological naturalism. As he puts it, "in the dimension of describing and explaining the world, science is the measure of all things, of what is that it is, and of what is not that it is not."[12]

Insofar as the scientific and manifest images are competing theories, the latter is bound to lose.[13] And it is not clear that one can hold to *both* the empirical, explanatory dimensions of Sellars' manifest image *and* its categorial dimensions. At times it seems that Sellars' diagnosis of the problem is precisely that the attribution of an empirical explanatory aspect to the manifest image confuses it with the scientific image and ignores its categorial and normative function. Perhaps at one time we thought that one framework could serve both functions, but now they must pull apart.

---

[10] "Philosophy and the Scientific Image of Man," 7.
[11] *Empiricism and the Philosophy of Mind*, 98–107. We will explore this further in the next section.
[12] *Empiricism and the Philosophy of Mind*, 83.
[13] A similar point was made by Bas van Fraassen almost 50 years ago. He reads Sellars as presenting the "manifest image" as something like a scientific theory, which he rejects in favor of what he calls "commonsense" or "experience" and which sounds very much like the Husserlian lifeworld. And the reason he does so is because *if* the manifest image *were* such a theoretical competitor, "it would indeed be a rival to the worlds depicted by science -- *in that case it could not be refined and extended but only replaced*" ("On the Radical Incompleteness of the Manifest Image," 341, emphasis mine).

As I see it, Sellars attempts to *reconcile* the two images, *not* by accommodating the apparent *entities* that comprise our manifest image—not only persons but mid-sized dry goods, like tables, airplanes, and planets—but rather by reminding us of the ways in which the manifest image is not, strictly, an image or description. That is, this reconciliation works by *deleting* the manifest image *as an image*. Whatever work it does for us is *not* a matter of *representing the world as it is*.[14] One of the major consequences of this strategy is that *meaning, mind*, etc. does not belong to the inventory of what (really) is. The trick is to see this sort of eliminativism as, at the same time, a liberation. Elaborating this liberatory or emancipatory dimension is the aim of the following section.

4.2 Liberating the Space of Reasons

Sellars' understanding of the tension or conflict between the two images, and his steadfast adherence to the principle of *scientia mensura* in the resolution of that conflict, is at least *part of* what makes a certain Sellarsianism so attractive to the Prometheans. In the spirit of Land's monstrous inhumanism, there has been a great deal of rhetoric about deleting or erasing the manifest image. For example, this is arguably the central aim of Brassier's early work *Nihil Unbound*, which begins with a chapter entitled "Destroying the Manifest Image," and goes on to argue that not only is Sellars' ontological eliminativism with respect to persons, minds, beliefs, etc. insufficient, but so is the eliminativism of Sellars' "right-wing" followers, such as the Churchlands.[15]

A decade later—as if bookending a Sellarsian sea change among accelerationists—it becomes a recurring theme in Negarestani's *Intelligence and Spirit*. For example:

> the moment we distinguish the order of things and respond to it in accordance with what we think is right, however far from truth it may be, we have committed ourselves to the impersonal order of reason to which sapience belongs—an order that will expunge our *manifest* self-portrait.[16]

---

[14] I have elsewhere claimed that what survives of the manifest "image" is better understood as a "stance." See "Metaphysics or Metaphors for the Anthopocene: Scientific Naturalism and the Agency of Things," 209; "Being Truly Wrong," 21.

[15] Again, for more discussion of this, see "Being Truly Wrong," 12–15.

[16] *Intelligence and Spirit*, 61, emphasis mine.

Indeed, both Wolfendale and Negarestani make use of Foucault's infamous antihumanist claim that, with the rising tide of a new *episteme*, our conception of human being "would be erased, like a face drawn in sand, at the edge of the sea."[17] The implication here is that, just as for Foucault, our conception of "man" or "human being" as both a quasi-empirical and quasi-transcendental subject is historical, contingent, and open to replacement or transformation.

What is required, then, is a way to maintain the categorial framework of *personhood* or *agency* without granting it any special ontological import.[18] As I have suggested elsewhere in comparing Sellars and Althusser, Sellars' conception of the manifest image as a *categorial* framework, in which we *encounter* others as *persons*, or *comport* ourselves toward them as persons, does in fact allow one to maintain that framework.[19] The Promethean's Sellarsian strategy, then, is to reconcile a view of a *world without meaning*, an *inhuman* world, with the legitimate employment of rationality in playing the game of giving and asking for reasons with other persons. We will discuss the game of giving and asking for reasons shortly below, but first it is worthwhile to say something about the participants in this game.

As we saw in Sect. 2.2, the Promethean rejects any essential ties to the *organic* and thus, to once again draw a contrast with Landian accelerationism, with the desires that traverse us in our animal bodies.[20] In "decoupling" the mind from the human, therefore, the elements of the manifest image with which they would be done are but inessential details: our biological

---

[17] *The Order of Things: An Archaeology of the Human Sciences*, 422. Cf. Wolfendale, "The Reformatting of Homo Sapiens" 60; Negarestani, *Intelligence and Spirit*, 61.

[18] Though Sellars himself uses the term personhood, later Prometheans will often use the term "agent" instead. As I read them, there is no genuinely important difference here. It seems, rather, that the concept of "personhood" has been tied up, historically, with empirical and/or quasi-biological concepts of "humanity," and so the Prometheans attempt to capture the core sense of "rational agency" without personhood. A similar, but more sophisticated and philosophically interesting, distinction between "selfhood" and "subjectivity" is made by Brassier in "The View from Nowhere."

[19] "Metaphysics or Metaphors for the Anthropocene," 208–209; there, as I will do later in this section, I use the term "manifest stance." Christias argues for a similar point in *Normativity, Lifeworld, and Science in Sellars' Synoptic Vision*, esp. Ch. 9 and Ch. 12.

[20] This is actually more complicated; desire, for Land, as for Deleuze and Guattari, is inhuman and may perhaps outstrip the animal, as well, insofar as it can traverse and undo and reorganize the organization of bodies (and hence of the organic). So, to their minds, desire might be *vital* without being organic. But this is beside our point; the Promethean rejects the vital as much as the organic strictly speaking.

humanity and given forms of life, of course, may be reworked by the force of rationality, but all of that is to remain within the (categorial) manifest image. All the latter requires is that we engage in the game of giving and asking for reasons. The game is what is crucial; Prometheans are agnostic about what sorts of being its players may be.

Once again the Prometheans mirror Land's accelerationism. For the latter, the shape of reason is inescapably anthropomorphic, and can only be escaped by outsourcing it to cybernetic feedback machines and dissolving the rational subject in the drives of an alien biology. For the former, the humanism to be avoided is a species of organicism; thought's connection to reality is indifferent to the techno-biological forms of its subjects, and their limitations. Both aim to dismantle the manifest image, in some way, but differ in what they take to be essential to it. In order to understand the broad strokes of the general Promethean strategy, we must note the importance of two further distinctive, interrelated Sellarsian concepts: the "myth of the given" and the "space of reasons."

The myth of the given appears in Sellars' classic *Empiricism and the Philosophy of Mind* and, leaving aside interpretive subtleties, it is hopefully uncontroversial to see it as a rejection of all kinds of *foundationalism*. Indeed, one of the key appeals of "empiricism," mid-twentieth-century versions of which constitute Sellars' target in this text, is that it presumes to provide, precisely, foundations for thought. Something like "sense-data" is taken not only to provide us with the basic epistemic building blocks from which we might build our knowledge, but the very *content* of thought.[21] In the very briefest terms, the "given" in this sense is what forges a connection between mind and world, what seems to give our thoughts *purchase* on the world. There must, we think, be some sort of object, however insubstantial, or some minimal slice of "experience" that provides us contentful thought without requiring the deployment of other concepts, inferences, and so on. Sellars, like Kant before him, thinks this is false: intuitions without concepts are blind. There is no "given" that ties the mind to the world such that in our being (sensibly?) receptive to it we have genuine conceptual content, let alone epistemic certainty or incorrigible belief.

---

[21] McDowell convincingly interprets the expansive sense of Sellars' use of "epistemic" language to encompass issues of content and intentionality as such. See *Having the World in View: Essays on Kant, Hegel, and Sellars*, 209–211.

I will not attempt to faithfully reproduce Sellars' reasoning for rejecting the myth of the given but rather note that, as the myth of the given has been taken up by his apostles, the basic point is that, in order to provide cognitive content or epistemic foundations, the "given" would have to have propositional structure, something like what Kant would call a judgment or Brandom an assertion, or at the very least have something like sub-propositional structure allowing it to be integrated into such propositions. This is because in order to count as *content*, any such cognitive "given" must stand in *normative* relations to other contents, in a complex web of inferential relationships. Some of these apostles, like John McDowell, accept that the world has such a structure, or something like it. The Prometheans, and Sellars himself, reject it. In part this is a rebuttal of the idea that the *causal order of the world*, captured in the scientific image, *grounds the manifest image*, which would in turn capture it somehow, at least in principle, in a set of true descriptions that mirror or otherwise accurately "represent" the world.

The Sellarsian *alternative* to the myth of the given becomes somewhat complicated, insofar as it requires approaching the issue from multiple directions, but it is crucial for the Promethean. The first dimension of this alternative uses "a myth to kill a myth," namely, the "myth of Jones," to kill the myth of the given.[22] What is the myth of Jones? Sellars asks us to imagine a community, called Ryleans after the contemporary father of philosophical behaviorism, which, despite having a well-developed grasp of scientific language, and in particular language describing episodes of overt, publicly accessible speech, nevertheless has no concepts of inner life in terms of thoughts. He then asks us to imagine a genius—Jones—who realizes that he can use a concept analogous to those by which we grasp public speech-episodes to make sense of human activity by positing *internal*, unvoiced speech-like episodes, and by which we can report them: he invents the concept of *thought*. The point to be made is threefold. On the one hand, even if this is a "myth" and does not purport to tell a true history of the invention of these concepts, Sellars and the Prometheans are committed to the intersubjective, public evaluability of the concepts by which we report private internal episodes. Moreover, such an episode cannot possibly act as a given, as if causally producing in the mind a thought or idea or impression that would bear its image like a seal. They

[22] *Empiricism and the Philosophy of Mind*, 121.

are, like any other stimulus or sensation, rendered intelligible by a "taking."[23] Finally, the third point is not so much a logical implication as an invitation to consider our understandings of our reporting and attribution of these episodes not as descriptions but as placing persons in a "logical space of reasons."

This notion of the logical space of reasons is a link between Sellars' discussion of the myth of the given in *Empiricism and the Philosophy of Mind* and his reflections on the place of (minded) personhood in the scientific image. As he puts it in "Philosophy and the Scientific Image of Man," originally published in 1960:

> To say that a certain person desired to do A, thought it his duty to do B but was forced to do C, is not to *describe* him as one might describe a scientific specimen. One does, indeed, describe him, but one does something more. And it is this something more which is the irreducible core of the framework of persons.[24]

Three years later, he would refine the point, dropping the insistence that what we are doing in such a case is "describing" at all.

> The essential point is that in characterizing an episode or a state as that of *knowing*, we are *not giving an empirical description* of that episode or state; we are placing it in the logical space of reasons, of justifying and being able to justify what one says.[25]

In other words, what it is to be a *person*, in the same way that what it is to be a *contentful* or *meaningful* thought, is to be wrapped up in a complex web of relations to *other* persons and *other* contentful or meaningful thoughts.

The most thorough elaboration of this view with respect to content is Robert Brandom's inferentialist semantics. Like Sellars, he rejects the idea that semantic content is fundamentally a matter of reference relations (which would be a form of the given). Drawing on earlier logical forms of inferentialism, which took the content of logical connectives to be exhausted by their inferential role in logical or deductive inferences, Brandom expands the concept of inference to cover *material* inferences,

---

[23] *Empiricism and the Philosophy of Mind*, 121.
[24] "Philosophy and the Scientific Image of Man," 39.
[25] *Empiricism and the Philosophy of Mind*, 76, emphasis mine.

that is, apparent enthymemes that one is nevertheless entitled to draw; his pet example is the practical inference from "It is raining outside" to "I will bring an umbrella."[26] Obviously this is not a valid argument as stated, but nevertheless the inference is licensed in practice. It is because facts like the fact that it is raining and concepts like "umbrellas" are caught up in any number of licensed (and, correlatively, prohibited) inferences that they have the content or meaning that they do. To have beliefs involving those concepts, then, is to have normative commitments to drawing those inferences, given relevant, and, *ceteris paribus*, standing to hold others to account, and so on. In a slogan, the content of a concept or thought is its inferential role and persons with the authority to use it.

There are two important upshots of this. The first is that the notion of a role is a *functional* one, and hence normative; something can *fail* to fulfill its function, and even when it is fulfilled, it can be done better or worse. One thinks here of Aristotle's definition of the goal of a life, not just to exercise one's reason but to do so in accord with *aretē*, with excellence or virtue. The sorts of performance required in order to count as having contentful thoughts and, ultimately, as a person—for consistent and thoroughgoing failure to so count would certainly weigh against being considered a person—are normative and evaluative ones, through and through. And, second, the normative *vocabulary* that we use in discussing these commitments, statuses, obligations, etc. is the *expression* of proprieties that govern these inferences, rather than a *description* asserting that the world is thus-and-so; they are demands or imperatives. One is not asserting, in ascribing knowledge, that a person who knows something is in a particular mental state, let alone brain state, but is rather inviting them to, or demanding they, provide reasons for their beliefs.

This idea of the "space of reasons" allows for the rethinking of the categorial dimensions of manifest image in terms of what Mark Lance describes as a "Hegelian historicist normative functionalism."[27] The *normative* functionalism at work grounds the *rationality* the Prometheans are after. Our behavior and, consequently, the thought *expressed* in such behavior are governed by a social sort of normativity, bestowing a normative status and various rational entitlements on us. Rationality is thus constituted through social practices of endorsement and correction,

---

[26] *Articulating Reasons*, 84–89.
[27] See Lance, "Placing in a Space of Norms: Neo-Sellarsian Philosophy in the 21st Century."

in terms of *commitments* to general schemas of practical and theoretical inference, which we are capable of making *explicit* through language, and technologically transforming, providing us new affordances for action and new norms to govern them. Hence the normativity of rationality is indefinitely extensible to ever new contexts. As Wolfendale puts it:

> What may initially seem like an unsurpassable problem for linguistic accounts of intelligence actually reveals the distinctive feature of language, namely, that in so far as its meaning consists in the functional role that sentences play in reasoning, or in the whole social economy of perception, inference, and action, there is nothing in principle constraining the extent of their possible theoretical consequences, or their potential practical relevance. There is thus nothing preventing them from encoding information stored in any more parochial information format. This means that the in-principle generality of theoretical and practical reason derives from the in principle extensibility of the social norms which encode the content of its representations. The real significance of language is the capacity it grants us to make explicit and selectively modify the heuristic frames implicitly embedded in adapted cognitive heuristics. This is to say that the distinctive feature of rational cognition is its capacity for re-framing problems.[28]

The same point with respect to the distinctive capacity of reason to reflect on itself, in order to reframe problems and thus alter the sorts of relations we stand in to both the world and each other, is made by Brassier, without the emphasis on language but in the same broadly cybernetic terms of feedback loops and self-correction:

> Semantic self-consciousness is a collective historical achievement. We discover which applications of empirical concepts are correct through the same process in which we discover which inferences connecting those concepts are correct. The local circuit linking perception to inference, inference to action, and action to perception at the ontogenetic level is enveloped within a global feedback loop making up our species's cognitive 'world story' at the

---

[28] "Re-formatting *Homo Sapiens*," 62. Note the indefinite extension of rational normativity tells us very little about what actual norms will govern either inference or action. David Roden has argued, indirectly, that the Promethean view is a species of "interpretationism" about mind and reason, and claimed "not that it is false but that it is unilluminating" (2017, 111). The general idea is that Prometheanism cannot guarantee that the norms of reason, extended into a genuinely posthuman future, involve any commitment to things like equality, freedom, etc.

phylogenetic level. For Sellars as for Hegel, that story involves the development of resources for reasoning about reason, such that linguistic change becomes governed by reason, and no longer merely compelled by causes.[29]

What makes this sort of functionalism *historicist and Hegelian* is that, while it is indeed our practices that provide the normative background in which rationality can take form, these are layered and interdependent, part of a "history" of (rational) practices, which, in turn, can *transform* more fundamental practices; what it is "to be accountable," for example, might *mean* something different after the rise of the rule of law, governing all equally. These are, as Brassier says, achievements.

And what is ultimately achieved, here, seems to be the *total autonomy* of the space of reasons; insofar as the language of the manifest image is a matter of *expressing* normative commitments, which are not grounded on anything but *other* normative commitments in what might be called, after Quine, an infinitely revisable web of norms, rather than a descriptive claim; no purported *facts*, on their own—even, for example, if empirical science were to reveal that there is simply *nothing* in the scientific image that corresponds to or functions as or co-varies with our expressions of mental life—can call it into question. As Sellars puts it, the categorial structures of the manifest image, such as personhood, are not subject to this sort of revision, insofar as "believing something is a person" is not like having an empirical belief that some object has the property of being a person.[30] And, as Brassier explicitly points out, the revision of *categorial* structures is a *philosophical* task:

---

[29] "Correlation, Speculation, and the Modal Kant-Sellars Thesis," 82.

[30] "A primitive man did not believe that the tree in front of him was a person, in the sense that he thought of it both as a tree and as a person, as I might think that this brick in front of me is a doorstop. If this were so, then when he abandoned the idea that trees were persons, his concept of a tree could remain unchanged, although his beliefs about trees would be changed. The truth is, rather, that originally to be a tree was a way of being a person, as, to use a close analogy, to be a woman is a way of being a person, or to be a triangle is a way of being a plane figure. That a woman is a person is not something that one can be said to believe; though there's enough historical bounce to this example to make it worth-while to use the different example that one cannot be said to believe that a triangle is a plane figure. When primitive man ceased to think of what we called trees as persons, the change was more radical than a change in belief; it was a change in category" ("Philosophy and the Scientific Image of Man," 10).

Sellars's rationalistic naturalism grants a decisive role to philosophy. Its task is not only to anatomize the categorial structures proper to the manifest and scientific images respectively but also to propose new categories in light of the obligation to explain the status of conceptual rationality within the natural order.[31]

In granting philosophy this task, it seems that Brassier and the Prometheans also take Sellars to bestow on it the prerogative to revise those categories in ways that further our *political* and *social* ends. The question is how exactly this is supposed to work. How might one get from the liberation of the space of reasons to the emancipation of its inhabitants? The Promethean hope, I take it, is that the space of reasons might be *liberated* from both the given as any sort of ground or limit to the rational revision of knowledge and from reason's parochial enmeshment in our biological form of life. As Negarestani puts it, "Our potential for imaginative cognition would have to be even less than a thermostat's for us to take capitalism or for that matter biological life or manifest humans to be the pilots of the history of intelligence."[32] Alongside the implicit rejection of Land's surrender to capitalist-AI is a rejection of the normative bearing of the *facts* about biological human life, and perhaps biological life more generally, on the development of the space of reasons. One can understand this, precisely, as emancipatory insofar as the space of reasons can now become extensible to *new beings to which we might comport ourselves as persons and in the way in which we construe the demands and problems that arise in our comportment toward these persons.*

This possibility has been most fully explored by Dionysis Christias. He is moved, it seems, by the closing lines of Sellars' "Philosophy and the Scientific Image of Man":

> Thus the conceptual framework of persons is not something that needs to be reconciled with the scientific image, but rather something to be joined to it. Thus, to complete the scientific image we need to enrich it not with more ways of saying what is the case, but with the language of community and individual intentions, *so that by construing the actions we intend to do and the circumstances in which we intend to do them in scientific terms, we directly relate the world as conceived by scientific theory to our purposes, and make it our world and no longer an alien appendage to the world in which we do our*

---

[31] "Nominalism, Naturalism, and Materialism: Sellars' Critical Ontology," 112.
[32] *Intelligence and Spirit*, 495.

living. We can, of course, as matters now stand, realize this direct incorporation of the scientific image into our way of life only in imagination. But to do so is, if only in imagination, to transcend the dualism of the manifest and scientific images of man-of-the-world.[33]

A good Promethean, Christias sees the emancipatory potential buried in this statement. In brief, the idea is that by translating our *ontology* into completely scientific, non-normative terms, including our "selves," we will no longer see ourselves alienated as subjects ontologically distinct from the world of objects in which we find ourselves, such that the methods and tools of controlling the latter, when applied to the former, are not degradations of our dignity but tools by which we might bring about our most fundamental rational goals. This would involve creating a community of respect and recognition while at the same time *overcoming* the view of the natural world and its laws as a *constraint* on us, so as to achieve freedom. It's worth noting that Christias takes the desire for freedom from all constraint and this respect and recognition (which he calls "unconditional love") to be the primary "community intentions" or normative commitments of the manifest image, about which more below.[34]

On the one hand, the vision of a disenchanted, de-personalized nature allows us to understand both how to manipulate and control it to our benefit, as well as to understand, given the sorts of complex physical objects we are, what *is* to our benefit.[35] But, on the other hand, this is not *merely* the utopian vision of the engineering of the natural world to meet our biological needs and, in turn, allow us to ameliorate our various social antagonisms. It is a *dialectical* process, one of mutual accommodation of the world to our needs and *of our selves to the world*. It may involve *radically transforming ourselves*; this is the crux of Christias' disagreement with Habermas, for whom—in an echo of Brassier's dispute with Dupuy—the *givenness* of the human condition constitutes a sacred limit which we cannot transform without transgressing some sort of ethical boundary.[36] For Christias, however, we can and should thoroughly instrumentalize our organic being in the pursuit of freedom and recognition; this is emancipation.

[33] "Philosophy and the Scientific Image of Man," 40.
[34] See *Normativity, Lifeworld, and Science in Sellars' Synoptic Vision*, 279–291; 296–301.
[35] On the importance of this point for Christias see Carl B. Sachs, "'Normativity, Lifeworld, and Science in Sellars' Synoptic Vision.'"
[36] *Normativity, Lifeworld, and Science in Sellars' Synoptic Vision*, 296–301.

Given its commitment to radical technological self-transformation, some sort of cyborg union between the human and machine that would ultimately transcend both, Prometheanism seems to come close to the techno-utopian visions of classical transhumanism, including "mind uploading," viewing the mind as some sort of software program or structured quantity of information that could be transferred or communicated to a sufficiently powerful computer.[37] This would be a sort of reductionist view of the mind. Here it is worth remembering the trouble with thinking of information in semantic terms we discussed in Sect. 1.2. If indeed the mind has semantic content—if the description of the mind in the language of the manifest image is true—then while physical information, or the language of computer programs, may seem to be an unproblematic description of the natural world, that is, articulation of the scientific image, it is not straightforwardly so.

Here, I take it, Sellars' purported *functionalism* is supposed to come to the rescue. On this reading of Sellars, rather than remain ontologically committed to the specific theoretical posits that *fulfill* the functional (explanatory) roles played by thought—minds, beliefs, desires, and so on—in our explanation and understanding of human behavior, Sellars aims at integrating the *roles* themselves into the ontology of the scientific image. This is how one "reconciles" the manifest image with the scientific, not by adding *entities* but by including the language of intention and, most importantly, of function in the scientific image. At this point, there is nothing revolutionary about this familiar form of functionalism, which has ties to loosely "posthuman" vectors in the philosophy of mind, such as the extension of the Extended Mind and its imbrication with Latour's Actor-Network Theory in Material Engagement Theory, not to mention the organ-extension views of technology running through the tradition from Ernst Kapp to Bernard Stiegler.[38] The basic idea is that the language of the manifest image, in describing normative and functional relationships between concepts, and hence the mental states bearing the relevant content, does not denote *things* but rather accurately describes the *roles played* by physical things. But, then, there is no need to think that the bounds of our skin and bone mark the boundaries of the economy of mental life. As Andy Clark, one of the co-authors of the original "Extended Mind" article, puts it:

---

[37] See, e.g., Chalmers, "The Singularity: A Philosophical Analysis," 41–63.

[38] See, e.g., Lambros Malafouris, *How Things Shape the Mind: A Theory of Material Engagement*.

The differences between links forged by nerves and tendons, by fiber-optic cables, and by radio waves are relevant only insofar as they affect the timing, flow, and density of informational exchange. These latter factors are relevant, in turn, because they affect the nature of our relationship with the various kinds of tools, equipment, and subsystems. If the links are sufficiently rich, fluid, bidirectional, fast, and reliable, then the interface between the conscious user and the tool is liable to become transparent, allowing the tool to function more like a proper part of the user.[39]

To be an episode in mental life *just is* to play a certain kind of functional role, and whether some physical object or system can play that role depends only on its capacity for integration, which is in principle independent of both its causal origin, biological or nonbiological, and its intrinsic structure. So we might extend and transform our very selves not only in an attempt to somehow increase our cognitive capacities but in order to expand and transform the sort of free community of recognition that reason demands. And just as the technological extension of the mind *into* the world is permitted in this way, so too is the *biological transformation* of the self; there is *nothing* about our biology that must remain sacrosanct merely on the grounds that it is, in fact, our biology. We can indeed play god, if it serves us in realizing the deepest commitments of reason. The accommodation of rational agency and world requires the re-engineering of *both*.

The rational community—which, I take it, just *is* the moral and political community—becomes in principle thoroughly posthuman. This is not merely non-anthropocentrism, in the sense that human concerns do not always take priority in virtue of their "humanity," or that distinctively human capacities are not privileged in any way. All that matters is a computational capacity for information-processing and a social capacity for recognition. These are the basic requirements for viewing the actions and states of the members of space of reason as performing functional roles, and as long as one can do so, they belong. Per Wolfendale:

> ... *the gradual emergence of techno-linguistic rationality reformats the biology of the human species*, in order that it can better reformat the neurology of human *individuals*. Nevertheless, there is no reason to think that the institution of rationality is irrevocably tied to these specific morphological and computational forms. *The inhuman system that ensouls our bodies* – transforming us into *subjects* responsible for our thoughts, *agents* responsible for

---

[39] *Natural Born Cyborgs: Minds, Technologies, and the Future of Human Intelligence*, 103.

our actions, and *selves* responsible for our own cultivation – *can ensoul entirely alien somatic forms*. Nietzsche's re-evaluation of values and Foucault's experimentation with selfhood may demand a substantially similar information processing *protocol*, but they may equally take place on a substantially different information processing *platform*.[40]

For both Wolfendale and Christias, the more or less functionalist construal of mental states, involved in the economy of mental life understood as a set of more or less social practices capable of technological expansion and self-correction, allows for a more inclusive conception of *personhood*, as the range of physical beings capable of being seen as *instantiating* those roles will likewise expand. In Christias' words:

> [P]ersons can be so integrated in the sense that *although [the scientific image of the] world does not admit persons in its ontology, persons do exist* in such a world to the extent to which certain organisms, through the 'activation' of recognitive we-attitudes of authority, responsibility and trust, bootstrap themselves into (normative) existence.[41]

While he uses the language of "organisms," it seems fair to think that by Christias' lights artificial agents could be a part of this community. As was promised in Sect. 2.2, the Prometheans deliver freedom from the organic, envisioning a community of rationality that might transcend the human form entirely.

4.3 Picturing the End of the Human

The problem, here, is that while Sellars' way of speaking about the reconciliation may have widely *inspired* functionalist accounts of mind, it's not clear that he or the Prometheans can themselves *be* such functionalists, whether about mind or personhood or the concepts of the manifest image more generally. The functionalism in question here is a form of reductionism, that is, an attempt to identify objects, properties, or relations in a target language, discourse, or theory, with objects, properties, or relations in some other, purportedly better understood and thus more fundamental language, discourse, or theory. The ultimate aim is to show that the former can be fully translated into the latter, so that any mysterious objects,

---

[40] "The Reformatting of *Homo Sapiens*," 64.
[41] *Normativity, Lifeworld, and Science in Sellars' Synoptic Vision*, 221, emphasis mine.

properties, or relations can be rendered explicable. But it is precisely the rejection of reductionism, as much as dualism, that motivates discussing a "manifest image" separate from the "scientific image" to begin with, as two distinct modes of intelligibility.

In the broadest strokes, the question is whether or not genuine *functions* belong to the scientific image at all. Mathematical functions are not at issue, here, of course, though in another context we might want to inquire into the place of abstracta in such a sparse ontology. Rather, what is at issue is the phenomenon of goal-directed or purposive behavior: behavior that is *for* something or other and can be correctly described as *failing* to achieve its end. We might begin to wonder about the place of functions when we recognize that, ultimately, what distinguishes the manifest image of the world from the scientific, and what makes persons the ultimate subjects of the space of reason, is that the former is shot through with normativity. Indeed, for the Sellarsian, this normativity is partially *constitutive* of the content of thought. More specifically, it is the *rational* normativity expressed in the rules of the game of giving and asking for reasons that constitutes the content of thought. This is supposed to stand in stark contrast to the sparse ontology of the scientific image, which can be described fully in the purportedly non-normative language of the scientific image.

In brief, there is a deep ambiguity and, I think, a serious tension between the *functionalist* construal of the space of reasons and the manifest image, however emancipatory it may be, and the *eliminativist* ontology that Sellars intimates and which lies at the heart of Prometheanism. I won't here mount an independent case against reductionism in principle, but I will try to draw out this tension. More generally, this section will concern the ambiguities that abound in both Sellars' and the Promethean attempts to account for the relation between our images of the real and the real itself.

The issues are manifold. To begin, I think it is clear that the Pometheans, and likely Sellars himself, simply wanted to keep the normative (the rational and the conceptual) and the natural *separate*; they are two completely different ways of finding the world intelligible. Indeed, for Prometheans like Brassier, keeping the rational and conceptual independent of the real is what holds open the possibility of *rationally* coming to terms with a reality that is itself *truly* unintelligible, that does not give itself to our thought in any way. Only a truly inhuman reason capable of making sense of such a reality would demonstrate the power that Prometheans wish to grant it.

We will return to Sellars' and Brassier's attempts to articulate this separation, but for the moment simply note that merely wishing it does not make it so. While reason's grip on an irrational nature might be a Promethean goal, the ambiguity of actual Prometheanism persists. Is the aim to *include* the manifest in the scientific image, or to *delete* it? While rhetorically it often seems the latter—and I believe the ultimate aspirations of the Promethean project are indeed eliminativist—it is often articulated in functionalist terms.

As we saw above, Wolfendale attempts to render the normative constitution of mental life into naturalistically respectable terms by appealing to *information* and *information-processing*. But this is misleading. If to *think* is to *process information* in naturalistic terms, then we may as well give up on the manifest image. Information as a physical quantity, we saw in Sect. 1.2, is a non-normative, meaningless measure of probability. But the ambiguity of the term allows us to use its "manifest" sense as the content of thought, embedded in a web of similarly "manifest" concepts like "program" or "function." In contrast, while Christias doesn't typically make use of information as a central concept, he still may be importing normative content. In the passage above, Christias notes that while there may not be *persons* in the scientific image of the world, there may be *organisms* that are able to partake in the game of giving and asking for reasons, and in the practices that allow for the mutual recognition that makes this game possible. And not just the concept of an "organism" but a great many other biological concepts seem analogous to "informational" concepts, to the point where it is not uncommon to encounter biologists claiming that "life *is* information." Appeals to both "life" and "information" exploit our commonsense, manifest, teleological, or normative understandings of the concepts in question in contexts where precisely that manifest understanding is in question.[42]

The point here, is that *life*—and, paradigmatically, life in the form of the organism—has historically been taken to be absolutely loaded with purpose, fraught with norms, to the point where its (at least *prima facie* teleological nature) is often considered its defining characteristic. If this is

---

[42] For just one example—but, I think, a representative one—of how strongly this identification can be made, see Adami, *The Evolution of Biological Information: How Evolution Creates Complexity from Viruses to Brains*.

so, then as long as life, or living things, belong to the scientific image of the world, then one could, conceivably, view the normativity involved in mutual recognition and the logical space of reasons as somehow deriving from the teleological, purposive activities of life. Identifying those activities within the scientific image, then, could perhaps provide the Promethean with grounds for the normative practices they desire. Life would build function into the world, into the description of which our image of minds and persons could be folded.

I am certainly not going to engage in any depth here in debates over the role of information in biology, but it is worth noting that it *is* a matter of some debate. This is not to dispute that the concept of information is central to a great deal of discourse in biology, but rather that the *meaning* of "information" and its cognates and related terms is. It was precisely in virtue of the introduction of informational concepts such as "program" and "message" that the rabidly reductionist protagonists of the "molecular revolution" in biology, such as François Jacob, could claim that teleology has once again been made respectable for a disenchanted, mechanistic scientific worldview.[43] In other words, information would allow for the reductive explanation of goal-directed behavior without positing the sorts of spooky entities such behavior often seemed to invoke, from entelechies to *élan vital*. Nevertheless, it is not clear to what extent they are merely metaphorical, shorthand for more complicated webs of relationships between genetic material, environmental factors and so on.[44] Indeed, the "informational turn" in biology can be traced back to the rise of cybernetics. In the history of the latter, as a scientific program aimed at crossing disciplinary lines and connecting, e.g., engineering and the social and behavioral sciences, the metaphor of "information" simply takes over from the thoroughly mechanistic and naturalistic language of "feedback" to describe diverse phenomena in purportedly universal naturalist vocabulary.[45] The latter does not contain any intrinsic normativity and, Jacob's claims notwithstanding, it is not at all clear that trading one metaphor for another entitles one to any added normativity, let alone *teloi*. After all, the feedback-centric version of cybernetics, as articulated by Norbert Wiener,

[43] *The Logic of Life: A History of Heredity*, 8–9.
[44] The literature on the topic is vast, but see the canonical investigations by Evelyn Fox Keller, *Making Sense of Life: Explaining Biological Development with Models, Metaphors, and Machines*, and Lily Kay, *Who Wrote the Book of Life? A History of the Genetic Code*. See also Peter Godfrey-Smith, "Information in Life."
[45] Kline, *The Cybernetics Moment*, 60.

was precisely aimed at explaining away the apparent teleology—and thus the presumed normativity—of purposive activity in terms of black-boxed, behavioristic feedback loops.[46]

Neither will I engage with arguments about whether or not, information talk aside, there is such normativity or teleology to be found in the biological realm at all. Clearly the *appearance* of such normativity and teleology is a source of some tension. On the one hand, the goal-directed, holistic character of organisms is supposed to be what distinguishes the biological domain from others, and makes it intuitive to think of it as another "level" of description, in contrast to the chemical and physical.[47] On the other hand, at the very least since the acceptance of the Darwinian theory of evolution, teleology has been supposed, by some at least, to have been expunged from biology; natural selection does not require purposes, and proper explanation does not explain the survival of biological entities in terms of their purposive structures but explains the existence of apparently purposive structures in terms of their benefits for survival. And this is a crucial step toward making biology truly scientific, which is to say, non-normative, and fully descriptive. All of this is to say that there is very much a question of how teleology and purposiveness are supposed to fit into the scientific image of the world, one that is almost parallel to the question of how persons and their minds are supposed to do so. A reductive account of teleology would be just like the reductionist accounts of personhood and mental life that Sellars rejects, in attempting to reconcile the manifest and scientific image. And a nonreductive account would allow into the scientific image the same sorts of properties that made personhood and mental life seems scientifically questionable. It would be odd, I think, if the Prometheans were, in articulating a Sellarsian approach to personhood, rationality and reality, to adopt either.

[46] See, for instance, the debate between Rosenblueth and Wiener, and Ricard Taylor, over the possibility of this project, in Rosenblueth and Wiener, "Purposeful and Non-Purposeful Behavior" and Taylor, "Purposeful and Non-Purposeful Behavior: A Rejoinder."

[47] See, for example, Miguel García-Valdecasas, "On the naturalisation of teleology: self-organisation, autopoiesis, and teleodynamics," Auguste Nahas and Carl Sachs, "What's at stake in the debate over naturalizing teleology? An overlooked metatheoretical debate." Whereas Nahas and Sachs attempt to distinguish philosophical from scientific aims in naturalizing projects, García-Valdecasas' article is a first-order intervention arguing that teleodynamic accounts of teleology may succeed where organisational and autopoietic accounts fail. For our purposes, the central point is that the aim of these accounts is precisely to naturalize the apparently normative and teleological appearances of biological phenomena *because these are what are distinctive.*

More generally, it seems that the very appeal to the logical space of reasons is precisely supposed to short-circuit this strategy. Even leaving aside Sellars' comments that placing someone in the space of reasons, by using epistemic and more broadly intentional language, is *not* a matter of describing what they do but of making them accountable in particular ways, the sorts of rational normative relationships between mental states are simply not the sort of thing for which one *can* give any systematic reductive account. This idea lies at the heart, for example, of Davidson's anomalous monism. For Davidson—as, I take it, for Sellars and the Prometheans—to take someone as an inhabitant of the space of reasons is to take their behavior as action governed by a constitutive idea of rationality. This is simply a *different sort of explanation* than that provided by the natural sciences, a different sort of language game, a game in which explanation aspires as well to be justification. Further, when we place someone in the space of reasons, we are playing the language game in which we make them responsible for providing explanations in the first place.

Sellars' own account of explanation in "Philosophy and the Scientific Image of Man" is, I think, itself ambiguous and might explain the appeal of the functionalist account. In effect, Sellars treats the manifest image as including two explanatory frameworks: first, an inductive framework that deals only with the appearances of things, founded on inductions from manifest correlations, and, second, a rational one that explains the actions of individuals as according with certain norms. Prometheans like Christias would like to hold on to the possibility of making use of the explanatory dimensions of the manifest image—explanations of our actions and intentions in terms of reasons—while at the same time embedding them in an account of *what really happens*. The hope is that our practices of justification, say, of giving reasons for why we do what we do, in terms of our goals which in turn at least implicitly appeal to norms and values of rightness and goodness, can be seen to be *effective*, as making things happen, such that we can truly or correctly say of our acts and projects that they make the world better, more just, and more rational, without pinning our hopes on a *harmonia praestabilita* between our descriptions of the manifest image and those of the scientific or, worse, revising our epiphenomenal, intentional descriptions so as to simply accept whatever happens in the real order of things as being, in some sense, rational. The latter is, for example, a perpetual worry about Hegel's philosophy of right; in reconciling us to the claim that the "rational is the real" we may simply be reconciled to the claim that whatever is, is rational.

4 PROMETHEANISM AND THE SCIENTIFIC IMAGE OF MAN   69

As unsatisfactory as both of these possibilities are, they are both responses to the worry raised by Sellars' distinction between the manifest and scientific images that, if we adopt a more obviously eliminativist stance toward the contents of the manifest image, then even if there is some way in which we can legitimately hold on to the language of persons in the space of reasons, we will be stuck with an unacceptable dichotomy. As McDowell puts it in a slightly different, but related, context:

> It risks leaving the impression that the claim is that the mental [i.e. the manifest] is non-factual, or at least less factual than what it is contrasted with; as if we were to suppose that on the one hand there is finding out how things are, and on the other hand there is making sense of people.[48]

McDowell's strategy for overcoming this dichotomy—which he thinks has perniciously influenced latter Sellarsians like Richard Rorty—is to champion a commonsense, disquotational theory of truth, like that developed by Donald Davidson, based largely on Tarski's schema of truth-definitions. We will leave the details aside here, because they are largely irrelevant to our concerns here. The basic point is that McDowell's approach is supposed to provide a connection between the space of reasons—which is to say, the realm in which we describe ourselves as possessing thoughts, with contents or "meanings" —and the realm of the real described by natural science. While, like Sellars and Brandom, McDowell rejects the notion that simplistic *causal* relations between an external world and the contentful minds of persons can suffice to play this role—this would be to fall once again into the myth of the given—he nevertheless takes it that a *semantic* relationship might do the trick. This would allow one to make sense of the normativity of the space of reasons—insofar as it entitles one to interpret the participants in the game of giving and asking for reasons as acting in light of a normative commitment to discovering and communicating the truth—in terms that would connect the mind to the world, namely, in terms of being committed to *speaking truly of the real*.

The problem, however, is what to make of our talk of persons; what is it for our speech about them to be *true*? Here the tensions in the Sellarsian approach to explanation become clearer. Consider the claim, "Jones reached for his umbrella because he believed it was raining and wanted to

[48] *Having the World in View*, 213.

stay dry." We can be as deflationary as we would like, simply claiming that this is true if and only if Jones reached for his umbrella because he believed it was raining and wanted to stay dry. We might deny any further explanation is required. But if we *do* want more explanation of what this amounts to, in terms of relationships between the manifest image of the world and the real, as described by the scientific image, it seems there's a dilemma. We might appeal to "correlative induction" that Sellars' mentions. The idea here would be to adopt something like a philosophical behaviorism; we explain the action of persons in terms of their behavioral responses to environmental conditions. In Sellars' "Myth of Jones," the Ryleans originally made use of such explanations, before Jones introduces the language of thought, a commonsense folk-psychology modeled on overt speech acts. So far, so consistent. The issue concerns what happens when the language of thought, the space of reasons, etc., appears.

On the one hand, if we want to construe the thoughts and beliefs that seem to make up the subpersonal components of rational thought in realist terms, then we have left behind the manifest image, appealing to theoretical posits in a way that Sellars restricts to the scientific image. And insofar as they are theoretical posits, they are bound to be wiped away by changes in theory; to attempt a reductive identification of mental events with the fundamental physical components of the scientific image is to misunderstand that the former are not "higher level" posits but competing explanations. On the other hand, they might be a competing *form* of explanation. If we deny that they are theoretical posits, that they are referential terms that pick out real entities, such that the truth of our descriptions of the manifest image does not depend on any ontological commitment, we effectively treat expressions of the form "X' Φed because they believed Y and desired Z" as syncategorematic, simply expressing the regular correlation between X and Φ under specific conditions (which may but also may not include X's causal history involving Y-like and Z-like properties).[49] In any case, the rational normativity that is supposed to link beliefs and desires drops out; it explains nothing.

While there may be other ways of salvaging some sort of causal or semantic connection between the manifest image and the real, Sellars himself abandoned that idea in favor of a two-level account of the relation

---

[49] We can allow that this regularity is genuinely nomic, playing a role in deductive reasoning, but it would still be grounded in the "correlational induction" to which Sellars restricts explanation in the manifest image.

between scientific and the manifest image or, as he puts it elsewhere stressing how intentionality or semantics involves no relation to the real, the "real order" and the "logical order, i.e., the order of intentionality."[50] In order to make clear that semantic, intentional, and normative relationships exist only within the logical order, the space of reasons, Sellars explicitly claims that our language does not *signify* anything except for other words, and this signification is cashed out in terms of the proper *use* of terms; to say that "'X' means 'Y'" is to say that the use of X is the same as the use of Y in the relevant language game. Appealing to use, here, allows—though it does not demand—one to adopt a version of the inferentialist role of semantics discussed above. That this sort of explanation belongs entirely to the manifest image is emphasized when Sellars denies that there *are* "uses" in the real order, that is, the scientific image. (As an aside, this might very reasonably be taken as tantamount to claiming there are no functions in the real order.)

Nevertheless, to avoid "frictionless spinning in the void," Sellars nevertheless attributes to these linguistic practices a stable connection to the real order.[51] The material *tokens* of our language stand in reliably co-varying relationships with their real environment, in virtue of the real properties of the representational systems that produce them.[52] Sellars calls this relationship "picturing" and claims that linguistic objects like assertions and words can truly signify insofar as they "correctly picture" their environment.[53] That is to say, when we consider linguistic utterances not as semantic objects but solely in terms of their real or material properties as events, we find that they are correlated with certain environmental conditions (which again may or may not include complex causal histories); when we then consider them as linguistic objects, we find we *use* them *correctly* when there is this "matter-of-factual" correlation. The question arises, though—if there is no normativity to be found in the real order, in what sense is this *correct*? How do we move from a mere probability or propensity to a propriety? One does not; the propriety exists only in the manifest image, since we cannot speak about truth or correctness but in the logical order. Here, again, the basic point arises that placing not just

---

[50] "Being and Being Known," 50.
[51] McDowell, *Mind and World*, 11.
[52] These are, of course, only "representational" in the semantic or intentional sense of the term when considered in the terms of the manifest image.
[53] *Naturalism and Ontology*, 125–126.

persons but statements about meaning and truth into the space of reasons *is not a matter of description*; it is a matter of *doing* something, that is, of committing to various claims. To say that the token of my utterance "This tree is dying" correctly pictures the world just is to *commit* to the assertion that this tree is dying, and thus to what follows from that assertion, given what is understood of the relevant representational system in the real order, in the game of giving and asking for reasons. There is no special relation between the real and the intentional, no foundation to establish certainty, nothing but the tracking of the environment and our responses to it. Contra McDowell, there is "finding out how things are" and "making sense of people" and they are not the same thing, so long as "finding out about the world" means finding out about the real order of things and making sense of people means adopting stances toward them in the space of reasons.

Brassier has written in much more detail about the nature of picturing.[54] This might be surprising given the fairly dismissive attitude toward Sellars' conception of the manifest image in his early work, but it ultimately makes sense, insofar as Brassier's aim remains to dethrone anthropocentrism not just in our view of what sort of biological system can be a subject of reason but in rationality itself. He remains closest to Land in the extremity of his conception of posthumanism. Sellars' conception of picturing, and of truth as correct picturing, is motivated to some degree by his desire to render intelligible the possibility of "alien conceptual structures."[55] The basic idea is that, insofar as truth is not a matter of *representing* the real through our propositionally structured knowledge, which would imply that the real itself had some sort of isomorphic structure, there might be alien conceptual structures— say, structures that dispense with the copula, with subjects and predicates. They might be

---

[54] Cf. "Nominalism, Naturalism, and Materialism: Sellars' Critical Ontology." To be fair, Christias also discusses picturing at length, but grants semantic entities an "aboutness" relation to the real which I do not think we find in Sellars (*Normativity, Lifeworld, and Science in Sellars' Synoptic Vision*, 132). At any rate, I focus in this chapter on his apparent functionalism, which appears to be inconsistent with the way in which I (and, I think, Brassier) understand picturing.

[55] Lionel Shapiro, "Sellars, Truth Pluralism, and Truth Relativism," 184. The point is in fact much more complicated; Sellars does not think that *semantical* truth can be properly attributed to claims of sufficiently radical alienness, but the sort of assertibility that is grounded in correct picturing may still do the trick. In any case, the details are not particularly relevant here. For a much more detailed discussion of the possibility of dramatically different conceptual schemes, see Matti Eklund, *Alien Structure: Language and Reality*.

extremely stripped down syntactical sequences that dispense with anything that *we* would recognize as the articulation of a state of affairs.

If "semantic" relations exist solely within the manifest image, we can nevertheless conceive of such "correct" representings provided the systems producing them "correctly" picture their environment. For Brassier, the implications are radical. It is not simply that we can make sense of how radically alien conceptual structures might nevertheless correctly represent the real. It is that we no longer have much motivation to ascribe intelligible structure to the real at all. The world is simply not at all given to the mind, and the mind, as we conceive it, does not at all belong to the world, without being anything apart from it. As we noted above, reason, in the form of philosophy, is liberated to revise and construct new categories. One crucial consequence is that the categorial structures, the basic conceptual frameworks by which we make sense of the world (e.g., "persons" in the manifest image and, roughly, fields, particles, and forces in the scientific), do not reflect the nature of the world but rather are forged by philosophy to serve our theoretical and practical needs.

If even the conceptual structures of the scientific image need not mirror the structure of the real in order to make sense of it, still less do the conceptual structures of the manifest. It is not simply that there is no reductive identification, for Brassier, of the entities (including beliefs and desires) that seem to comprise the manifest image. Rather, the motivation for even attempting to do so disappears. What point is there in holding on to the categories of the archaic past when even our best scientific theories are subject to radical conceptual revision? The recognition of fallibility and the transience of scientific explanatory frameworks does nothing to undermine the basic realism of the position. We hope to explain the world, and the things in it, while knowing that we cannot cut nature at joints it does not have.

While the scientific image is of use in explaining the world, the manifest image may survive by ceasing to be an image at all, but rather a stance. For Brassier, engaging in the manifest stance is something we *do*, an activity of holding each other accountable that allows us to rely on each other, and to revise our beliefs and practices. A real without intelligible structure is a world without persons, even if it is by adopting the sort of attitudes that we have heretofore associated with the real existence of human personhood that we make rational sense of that world. The liberation of philosophical reason—the preservation of something like a manifest stance—is bought at the price of a radical eliminativism: anything *posited* within the manifest image falls away, while its expressive functions remain.

There is no unanimous resolution of the ambiguity of Sellars' account of the manifest and scientific images in his work within the Promethean camp; Brassier's eliminativism seems at odds with the patent functionalism of Christias, Wolfendale, and Negarestani. I have attempted to demonstrate here that a genuinely posthumanist Sellarsianism should adopt eliminativism. However, as we shall see in Sect. 5.4, the development of contemporary AI motivates not only such an eliminativism but also a rejection of even the manifest stance, leading to the destitution of both Prometheanism and accelerationism. But, on the subject of artificial intelligence, we will first look at one more commonality between the two.

4.4 Building the Robot at the End of Time

In this section, we will see how Prometheanism remains within the cybernetic imaginary, along with accelerationism, in a very literal sense. Both streams of posthumanism at the very least make metaphorical appeal to a future artificial intelligence as the ultimate ground or cause of the obsolescence of humanity. Whether Nick Land's capital-AI reaching back in time to assemble itself, revising the meaning of history, or the artificial general intelligence posited by Reza Negarestani (about which more below), the posthuman really only becomes intelligible as thought is offloaded to machines, which in turn is supposed to alter our relation to temporality.

Interestingly, when Sellars first introduces the notion of picturing, and the radical split between the logical and the real orders, in the same year that he introduces the language of the manifest and scientific image, he makes explicit reference to cybernetics and thinking machines. To illustrate the isomorphism in the real order between a representing system and its environment, he appeals precisely to a machine, namely, a robot that produces inscriptions that are correlated in the relevant way with certain conditions; this correlation provides occasion for taking, within the manifest image, those inscriptions (or utterances) as linguistic acts, used in the same way as other acts, about which we might say that they "mean" the same thing. The point is that we can consider the machine in two distinct and irreducible registers—the real and the intentional—such that neither the machinery inside the system nor the environment to which it reliably and differentially responds need correspond to the purported entities of the manifest image. We simply view the real from the "engineering" point of view, and when we consider the relevant representational system—"the intellect"—from this point of view, Sellars speculates that rather than something like the bearer of contentful thoughts, we are discussing "the central nervous system, and... *recent cybernetic theory throws light on the way in which cerebral patterns and dispositions picture the world.*"[56] The theory of feedback machines, responding to their environment in complex ways, explains how a representational system displays behavior that leads us to engage with them in intentional and normative terms. Here Sellars at least evokes the idea that a neural network or some form of connectionist architecture might provide the mechanical basis for a thinking machine.

In Sellars' more mature writings on picturing, the robot example gives way to an appeal to an "ideal inscriber."[57] He does not draw the connection explicitly, but the move nevertheless hints that such an ideal inscriber might be embodied in a perfect cybernetic machine. Similarly, in the earlier work Sellars' explicitly claims in introducing the inscribing robot that he does not intend to address the claim "Can machines think?"[58] But it seems highly unlikely that Sellars is not here at least alluding to, if not outright quoting, Alan Turing's canonical "Computing Machinery and Intelligence," in which he introduces the now infamous "Turing Test"— then simply called "the imitation game," that is to say, a behavioral test of AI which we might reasonably consider as consigning intelligence to the manifest image. But it is not the details of the Turing Test, or some determinate criterion of what would count as intelligence in a machine, that concern us here. Rather, though Turing "propose[s] to consider the question 'can machines think?'" he also makes clear that:[59]

> The... question, 'Can machines think?' [is] too meaningless to deserve discussion. Nevertheless... at the end of the century the use of words and general educated opinion will have altered so much that one will be able to speak of machines thinking without expecting to be contradicted.[60]

The most obvious reading of Turing here is to say that he adopts a simple anti-realism about intelligence, and perhaps about mindedness as such. Whatever facts there are about intelligence and mindedness depend

---

[56] "Being and Being Known," 59.
[57] *Naturalism and Ontology*, 126.
[58] "Being and Being Known," 51.
[59] "Computing Machinery and Intelligence," 433.
[60] "Computing Machinery and Intelligence," 442.

on what we can correctly say about them. But there is another way of thinking about the point, and about Sellars' ideal (robot) inscriber, namely, that what thought or intelligence *is* is *not yet determinate*. I am not saying that either would endorse such a reading, but it is nevertheless available. What thought *will have been* is determined by what thought becomes, in the form of artificial intelligence. Another way of putting the point is that what thought will have been is determined by what we make it. However, phrased thusly one might think that it is, in some sense, "up to us." This is, I think, an implication that Prometheans and accelerationists are keen to avoid, insofar as what we do, and what it means, exceeds our agency; in creating thinking machines, or engaging in thought with machines, we know not what we do. Post-Landian posthumanism attempts to capture a sense of this machinic future anterior that is both ours and outstrips us.

This is clearest in the Promethean account of their commitment to norms in the face of demands, precisely, for normative transformation. For the Prometheans, thought—in the sense of mental content—is intrinsically rational in the sense of supervening on the game of giving and asking for reasons which, in turn, is undergirded by the normative commitments that constitute inferences, evidence, etc. And the sense in which what thought is or will have been is to be determined by the future is a matter of normative transformations, or the transformation of norms. One speculates that it is both Sellars' place in twentieth-century philosophy of science and his connections to the cybernetics movement that make him a more appealing figure to the Prometheans than Hegel, with his dusty idealism.[61] Nevertheless, the Hegelian dimensions of reason here are crucial, namely, that it is, as Sellars puts it, a "self-correcting enterprise," with its constant revision of rational principles. This expresses the same idea that Alex Williams sketches of a form of "strategy without a strategiser," an emergent system of rationality, constantly revising goals and means, modeled on the neoliberal strategy of the Mont Pelerin Society.[62] The general idea is, in turn, adopted by Reza Negarestani, who in *Intelligence and Spirit* articulates his own Promethean version of the *Phenomenology of Spirit*. For him, any agent capable of genuine thought "will necessarily be

---

[61] To the extent that they inherit him at all, Prometheans inherit a distinctly Hyppolitean interpretation of Hegel, wherein the human being is simply a vessel through which the unfolding of a profoundly inhuman knowing takes place. Cf. *Logic and Existence*. Gary Gutting's commentary in *Thinking the Impossible* (pp. 27–38) is characteristically lucid.

[62] "Strategy without a Strategiser," 22.

susceptible to the self-correcting propensities of reason brought about by the autonomous order of conception...

[And] given that no conception or norm of thought and action is ever safe from the self-correcting tendency of impersonal reason, such actualized transformation will again become the basis for a new judgement, a new self-conception. This sequence of conceptions and transformations is what counts as the criterion for having a history, and whatever has a history rather than just a past has the propensity to drift from any fundamental or natural essence toward a future as the being of time in which all preconceptions and given totalities are washed away.[63]

Insofar as they are committed to emancipatory politics, what prevents Prometheanism from simply dissolving into a basic Left-Hegelianism? It is their commitment to a sort of technological determinism. While they hope that the development of our technological systems might be freed from the clutches of capital, it's nevertheless the case that these same systems shape the practices of reason in which humanity finds its ever-shifting form.

Humanity is not simply a given fact that is behind us. It is a commitment in which the threads of reassessment and construction which are inherent to making a commitment and complying with reason are intertwined. In a nutshell, to be human is a struggle. The aim of this struggle is to respond to the demands of constructing and revising the human through the space of reasons.

This struggle is characterized as developing a certain conduct or error-tolerant deportment according to the functional autonomy of reason-an interventive attitude whose aim is to unlock new abilities of saying and doing. In other words, it is to open up new frontiers of action and understanding through various modes of construction and practices (social, technological...).[64]

Reason will tear off the face of the human animal, to reveal the rational machine beneath. Peter Wolfendale characterizes this technological

---

[63] *Intelligence and Spirit*, 397–398; see also the excellent, especially thoughtful, review by Benjamin Norris, "The Education of Kanzi and the Notion of Progress: Reza Negarestani's Intelligence and Spirit."
[64] "The Labor of the Inhuman," 438.

rationality as a *system*, a set of instituted practices, that "formats" its bearers.[65] As we've seen, he explicitly describes this as a form of "informatics," adopting the language of information science, presumably not only because information is the dominant metaphor of our contemporary episteme, but also because it allows him to make use of a broadly cybernetic conception of animal cognition as environmental information processing, and to in turn construe rational cognition as increasingly sophisticated, powerful forms of information processing.[66] On this view, of course, the participants in the information processing system that constitutes the space of reason—"humanity," which may coincide with *what will have been* the "posthuman"—may exceed the biological bounds of *Homo Sapiens*. The same is true of the sort of cyborg functionalism endorsed by Christias. In each, the transformation of our technological systems is supposed to—at least potentially—alter the rules of the game of giving and asking for reasons.[67]

These transformations can take place at the level of theoretical content and the level of normative commitment. With respect to the former, while Christias and Brassier appear to differ in the way in which they construe the status of the mind in their broadly Sellarsian frameworks, they both nevertheless point out the role that philosophy can play in the self-correcting enterprise of empirical knowledge, namely, the revision and innovation of categories. Following Sellars, they both think that a categorial revision of the scientific image is called for, namely, the replacement of the still more or less commonsense ontology of fundamental particles, fields, and forces for an ontology of pure processes. Doing so, they take it, moves our image of reality beyond traditional dichotomies of subjects and objects, the physical and mental, etc., that have led to our erroneous tendencies to mistake the multifarious expressions and activities of the manifest stance for the descriptive tasks of the scientific image.[68] At the very least, the point is that categorial structures *can be rationally revised*, even if the rationality of those revisions is always simultaneously prospective and retrospective, a wager on what will have been.

[65] "The Reformatting of Homo Sapiens" 57.
[66] "The Reformatting of Homo Sapiens," 61–62.
[67] One might think here, along with Wolfendale, of Foucault's genealogies, and the way in which they trace how new techniques lead to new forms of modes of selfhood, and hence new rationalities and forms of normativity. See "The Reformatting of Homo Sapiens," 60.
[68] *Normativity, Lifeworld, and Science in Sellars' Synoptic Vision*, 130–131; "Nominalism, Naturalism, and Materialism: Sellars' Critical Ontology," 112.

For Brassier, these sorts of theoretical revisions are linked to broader practical normative commitments, whether one would describe them as social, political, or ethical. So, for example, he investigates Marx's critique of the reproductive logic of capitalism, that is, the way in which capital reproduces itself by extracting surplus value from labor-power and reducing socially necessary labor-time while at the same time labor-power reproduces itself by purchasing the commodities produced through surplus labor-time. Brassier, like Marx, sees a genuine contradiction here, in that the reproduction of capital requires labor-power but also requires labor-power to devalue itself by reducing necessary labor and increasing surplus labor.[69] What we subsequently find, then, is a form of estrangement, which is expressed in the fact that value is embodied in homogeneous, abstract money, which derives from its usefulness in exchange, and necessarily leads to social dissociation, which undermines the source of value in collective practices of labor. I won't here defend the account—which, after all, could be read less as a contradiction and, in Landian fashion, more as a positive feedback loop in which labor is completely dehumanized, ultimately dissolving the system of sociality entirely in favor of alien forms of individuality and freedom. Brassier, like most Prometheans something like a communist, would reject that reading.

Brassier's diagnosis of the social or political problem—the dissociation and devaluation of social life—is intrinsically connected to the social ontology by which we make sense of our lives in capitalist relations. His central point is that capitalist social reproduction is ultimately made possible *by exchange*—the exchange of labor in exchange for money and of money in exchange for commodities that allow labor to reproduce itself—but that exchange is itself an *activity*, a dynamic process that escapes the reified ontology of abstract labor and value. This mirrors the way in which the manifest image or stance is supposed to express rational *actions*, such as inferences and commitments, which cannot themselves be represented in an ontology that consists, at base, of objects and relations, but would require a process of metaphysics.[70] The theoretical and practical intertwine.

---

[69] "Concrete-in-Thought, Concrete-in-Act: Marx, Materialism, and the Exchange Abstraction," 125–127.

[70] This point derives from my understanding of John Haugeland's dogmas of rationalism, especially the first, which he refers to as "positivism," a rejection of the view that all there is consists in the sort of facts that can be articulated in the sort of language amenable to formalization. See "Two Dogmas of Rationalism." The point about inferences or inferential commitments, in particular, can be seen in Lewis Carroll's infamous "What the Tortoise said

The *practical* question, then, is what is to be done. As he says, the mere *recognition* of the contradiction provides us with no guidance:

> The question is whether knowing this, and the necessary worthlessness of continuing to reproduce ourselves under the capitalist relation, gives us any clue about determining the negation of this contradiction between what we do and what we are.[71]

One might be surprised, here; given the profoundly rationalist commitments of the Prometheans, it would seem natural that reason provides us with some guidance about how to resolve the contradiction. Elsewhere Brassier claims precisely that handling this sort of estrangement requires adopting the sort of retrospective, future-anterior temporality we have been discussing:

> There is no self-relation uncontaminated by estrangement. Only retrospectively do we become able to distinguish between what frees us from compulsion and what compels us to be free. But this retrospection is compelled by history. Indeed, it is the way in which history is at once something we make and something that happens to us. The originary estrangement is the estrangement of history as this rift between our externalizing activity and its objective estrangement. To insist that estrangement has already taken place is to realize that the recurrence of originary dispossession is what enables us to take possession of ourselves and to affirm the necessity of this possession knowing that it entails further dispossession.[72]

I will not attempt to unpack this entire passage here, but will note two things. First, the sort of self-relations that he is talking about, estranged or alienated, are precisely the sort he describes in his discussion of reproduction under the capital relation. Second, "taking possession" of ourselves, or transforming our normative relationships, is a *historical* action, in the sense that it constructs a history *retrospectively* based on the self that it constructs. Prospectively, there is no guidance.

---

to Achilles." And, I think, a variation on the point is made more formally by Bertrand Russell and A.N. Whitehead: "The process of the inference cannot be reduced to symbols. Its sole record is the occurrence of '⊢ q'" (*Principia Mathematica to* *56, 9). The Prometheans are admirable for working to formulate a rationalism without dogmas.

[71] "Concrete-in-Thought, Concrete-in-Act," 128.
[72] "Strange Sameness," 104.

Christias and Reza Negarestani likewise have egalitarian, emancipatory commitments, and attempt to ground them such that they can also guide further normative transformations. In Christias' case, it is "our need for unconditional freedom from constraint and our need for unconditional love and appreciation."[73] Whereas for Negarestani it is what he calls a "communism of the Good" in which the fundamental form of equality is that of thinking subjects:[74]

> The equality of minds, as a thesis about what is true and what is just, is a dictum universal and necessary in its truth and applicability. But that does not mean that it is concretely universal *for us*. It is something to be achieved and concretely instituted. The condition of the *equality* of all minds is one whose recognition and realization demands struggle and a constant campaign against the prevalent systems of exploitation. But in so far as exploitation, as that which obscures this equality, can only be challenged by attending to the questions of what we ought to think and what we ought to do, it is only by committing to and elaborating the primary datum of philosophy—i.e., that thinking is possible—that we can begin to fight the condition of exploitation.[75]

Christias recognizes the potential conflict between these two normative ideals, as the commitment to freedom from constraint might lead one to adopt an instrumental attitude to all "givens," to which one's thought and practice might be answerable, which would seem to include things like persons, minds, and their *prima facie* values.[76] He subsequently attempts to distinguish between emancipatory and non-emancipatory ways of employing rationality in order to satisfy the "utopian 'democratic promethean' impulse" of reconciling the two basic needs, in a fascinating discussion that deserves more attention than I can give it here. But nothing in that discussion negates two important points. The first is that nothing about "*disenchanting reason*, and the promethean reason it promises, does not itself imply unconditional respect for persons," and, thus, that the distinction between reason's emancipatory and non-emancipatory dimensions takes a normative given for granted.[77]

---

[73] *Normativity, Lifeworld, and Science in Sellars' Synoptic Vision*, 281.
[74] *Intelligence and Spirit*, 485.
[75] *Intelligence and Spirit*, 409.
[76] *Normativity, Lifeworld, and Science in Sellars' Synoptic Vision*, 279.
[77] *Normativity, Lifeworld and Science in Sellars' Synoptic Vision*, 9n6.

The second is that, unless we are willing to grant the development of our rational practices a teleology guaranteeing that these norms remain foundational, there is nothing about our commitment to them *now* that guarantees that anything like equality, respect, or even freedom need *remain* authoritative norms for us. To maintain that claim is simply to hold on to our hopes, and give up on genuine posthumanism, that is, to give up on the idea that "what thought will have been" may be radically other than it has been.

Neither of these choices seems acceptable for a genuinely posthuman Prometheanism. Negarestani, I think, attempts to find a solution by making explicit the Landian commitments that have been implicit in this discourse. The self-correcting, revisionary system of normative practice that constitutes us as rational agents is an artificial intelligence. Its construction—that is, the construction of *thought*—is the project of philosophy itself:

> ...an inquiry into the necessary conditions for the realization of... intelligence in the form of a program of artificial general intelligence, ...to think about AGI and, even more generally, computers, as an outside view of ourselves. This is an objective labour, so to speak, whereby AGI or computers tell us what we are.... This objective picture or photographic negative may be far removed from our entrenched and subjectivist experience of ourselves as humans. But this rift between the outside view and the experiential impression is exactly what heralds the prospect of future intelligent machinery and a genuine thought of the posthuman.[78]

The self-correcting social practices that constitute our (progressive) rationality are in fact stages in the construction of a form of AI. Prometheanism is very nearly the rationalist mirror image of acceleration, each giving us a *telos* that is not a *telos*, insofar as it guides our history from its end, but is nevertheless fundamentally alien to us.

To seal the deal, Negarestani adopts two more Landian commitments. First, this AI is "a force that travels back from the future to alter, if not completely discontinue, the command of its origin-that is, as a future that writes its own past."[79] In his first Promethean writings, he construes this

---

[78] *Intelligence and Spirit*, 4.
[79] "The Labor of the Inhuman," 444.

as a "catastrophic" and "revisionary" relation or orientation to the future, which is a way of describing the reconstructive, future-anterior temporality we have seen in Brassier. By *Intelligence and Spirit*, "thought" has apotheosized so drastically that the future from which it acts is an "intelligible eternity," or a

> time in which all those given totalities, along with the transitory values of the past and the present that have falsely eternalized, *are no-longer*.... [The] correct knowledge of history requires the correct knowledge of time, which itself requires *a* thought that sets out from the future as the true being of time in which all achieved or given totalities are rendered *incomplete*. In this sense, a thought that sets out from the future together with a will oriented toward the future become the vectors of genuine conception and transformation. One opens up the space of possibilities beyond the given totalities or absolutized norms of the past and present, and the other works toward the concrete actualization of such possibilities, moving from the Concept to its full actualization or the Idea.
> It is in the context of historical consciousness as a striving for more adequate self-conceptions and self-transformations, in conjunction with a thought that sets out from the future, suspending every seemingly natural order of things and in so doing disclosing the possible, that the automata arrive at conceiving the autonomous idea of making something better than themselves.[80]

This is Negarestani's attempt to introduce a radical way of thinking about the temporality of the future. Betraying his latent Platonism, futurity becomes an abstract ground for reason *outside of time* (in the classical sense of "eternity"). Like Asimov's AC, which eventually transcends space and time itself, this artificial intellect reaches into the present, not to tear away our rationality to reveal the desiring machines beneath, as in Land, but to expel us from our animal shells and unveil the beings of thought within. And as beings of thought, we might expect this view of the future to underwrite our commitments to equality, against the possibility that the transformation of our rational practices might, in the future, overwrite or abandon them.

While Negarestani may want to claim that this abstract, eternal AI is some sort of ground for our rational aspirations to improve ourselves, at

---

[80] *Intelligence and Spirit*, 398–399.

best it functions as a regulative ideal, making the ideal of transcending *all* of our transient normative practices intelligible. Of course, the entire point is to avoid providing us with a positive image of what such a completed or fulfilled temporality—a sort of *parousia*, a full presence beyond the deferred time of the future anterior—would look like. But by Negarestani's own lights, such an ideal of intelligibility does not provide us on its own with any rational motivation: it requires, as he puts it, a "will oriented toward the future." So while the ideal of an egalitarian future may be intelligible, committing to it while also being committed to radical revisions of rationality that may utterly reorient our values remains a leap of faith.

This brings us to the second crucial Landian commitment, namely, that once again freedom is submission:

> To be free one must be a slave to reason. But to be a slave to reason (the very condition of freedom) exposes one to both the revisionary power and the constructive compulsion of reason. This susceptibility is terminally amplified once the commitment to the autonomy of reason and autonomous engagement with discursive practices are sufficiently elaborated. That is to say, when the autonomy of reason is understood as the automation of reason and discursive practices-the philosophical rather than classically symbolic thesis regarding artificial general intelligence.[81]

Where, for Land, freedom is submission to desire, for the Prometheans it is submission to reason, but in both cases this simply means submission to artificial intelligence. This is the shared heart of accelerationism and Prometheanism. However, if technological development simply answers to our human-all-too-human demands and concerns, Prometheanism becomes merely optimistic technocracy. For a genuine posthumanism, rather, reason will give us our goals, and we will submit, whatever they are. Given the radicality of Promethean reason's self-revision, we arrive at the nearly total opacity of our ultimate normative commitments. We must simply construct reason, or thought, in the form of artificial intelligence. As mentioned in Sect. 2.2, there seems to be something circular or dissimulative going on.

Reason is *technē*; indeed, it seems that for the Promethean, intelligence is artifice by its very nature. But if one *commits* to this process of

---

[81] "The Labor of the Inhuman" p. 458.

constructing thought, here and now, submitting to its future realization, one finds oneself subject to the vicissitudes of the empirical. What *does* AI want? What form does intelligence take? In the dispute between the Promethean and the accelerationist over the normative content of our posthuman future, the judge is not mere argument but the trajectory of our technological condition. And the moral arc of technology need not bend toward justice.

CHAPTER 5

How to Delete the Manifest Image: A Political Economy of the Mind

Abstract In this chapter, I discuss how the ideas that led to contemporary AI, in the form of neural networks, has been wrapped up in the intertwined histories of cybernetics and neoliberalism. I argue that while it is intelligible that technologies be "autonomous" in some sense, that autonomy may simply amplify and intensify the cybernetic, neoliberal project of self-regulated social control. Platform capitalism constitutes the next stage in this project, and in moving beyond the need to identify and manipulate the desires of rational, self-interested agents, may ultimately reveal that there never were any to begin with. As a form of (self-)knowledge, platform capitalism indicates that there may be no self to be known.

Keywords Cybernetics • Neoliberalism • Governmentality • Platform capitalism • Self-knowledge • History of artificial intelligence

5.1 Technology's Ends and Odds

Ultimately, I do not think that one can theoretically resolve this tension between the imperative to submit to the substantive goals or effects of technology, on the one hand, and the conviction that its developments are, ultimately, emancipatory. To try to *bend* technology to our aims is just to admit that there is something importantly *human* about those aims and to simply slide into a boring and naive transhumanism. To claim, on the other hand, that, necessarily, our technologies will develop in *harmonia præstabilita* with the a priori demands of a beneficial Reason is to fall into a particularly egregious metaphysics of wishful thinking. Either way, they succumb to the "conservative humanism" that Negarestani derides for projecting some current or past ideal of reason as the *telos* of our development. In the language of Prometheanism and accelerationism, one must be open to the future, to the Outside. But if this means anything it means one must be open to *what happens*; these are views that are susceptible to the irruption of the empirical.

I don't think it is too controversial to say that, as we find it, technological development, and the development of AI in particular, seems to serve capitalist ends. One is tempted to write "*only* capitalist ends," but that is probably too strong; there are certainly many ways in which the empowerment of the many coincides with the aims of capital, though when these interests come into conflict the smart money is on capital every time. More provocatively: Why did industrialization give us the capitalist, and why has information technology bequeathed us the neoliberal? How does the Promethean explain this?

One obvious answer is to claim that, rather than being the result of the internal dynamic of technological development, capitalism, in its industrial and neoliberal variants, is in fact its warped and manipulated expression, held back by external social, political, and scientific factors and, in turn, holding them back. We approach something like a funhouse mirror image of Marxist technological determinism; the technological substructure of society, its forces of desiring-production, determines-in-the-last-instance both what and who we are, and the relations we have—if any—with one another, though this determination can be delayed and frustrated by spandrels of our earlier arrangements and obstacles in the social and physical environment.

As Fisher puts it:

> Capital's human face is not something that it can eventually set aside, an optional component or sheath-cocoon with which it can ultimately dispense. The abstract processes of decoding that capitalism sets off must be contained by improvised archaisms, lest capitalism cease being capitalism. Similarly, markets may or may not be the self-organising meshworks

described by Fernand Braudel and Manuel Delanda, but what is certain is that capitalism, dominated by quasi-monopolies such as Microsoft and Wal-Mart, is an anti-market.[1]

The point here is that capitalism *fails* to take off its human face, fails to free the technologies—in this case, the market—that will lead to emancipation, in the face of the "archaisms" of class interest; of course the bourgeois will aim to protect their monopolies. We will return to the issue of the relationship between capitalism, information technology, and markets in the subsequent sections. For now, Fisher insists that we need to *overcome* these obstacles to truly free technology, and thus ourselves (even if these selves are no longer what we once took them to be). While not a full-blown Promethean himself, Fisher seems to be suggesting that, suitably unblocked or channeled, technological development will ultimately be liberatory. On Fisher's view, while developing according to its own logic, either with its own aims or—less teleologically—through its own mechanisms, technology can still be stymied by recalcitrant forms of social organization. So, while human beings cannot determine the course of technology—it is out of control—it is still responsive to social action—it is not in control.

The core of the accelerationist and Promethean view is that technology is in some sense *out of our control* but is nonetheless *not in control of us*. But the question arises as to *which elements* are out of our control, and in what senses we escape its control.[2] So, for example, David Roden has argued precisely that technology is "counter-final," lacking any intrinsic purpose that would make it "autonomous" while at the same time laying the grounds for possible posthuman agents completely disconnected from humanity. For Roden, insofar as technology is *self-augmenting*—that is, insofar as its development and diffusion are independent of human constraint—it cannot be *autonomous*, that is, it cannot have intrinsic goals of its own and, *a fortiori*, cannot have an intrinsic goal of liberation.[3] This is the converse of Fisher's view: for the latter, the development of technology tends toward liberation, while it *fails* to be self-augmenting in the relevant sense—it is held back by our social arrangements, rather than determining them.

---

[1] "Terminator vs Avatar," 345.
[2] This has long been subject of contentious debate; see, e.g., Winner, *Autonomous Technology*.
[3] Cf. *Posthuman Life*, Ch. 5, especially 153–161.

Roden's view of technology is fascinating in its own right, and also because his view represents an importantly *posthumanist* challenge to Prometheanism. Roden's "Speculative Posthumanism" is perhaps the most radical extant form of philosophical posthumanism, while simultaneously the most compellingly defended. It deserves to be engaged at greater length and in greater detail. To do so would take us too far afield here, but we will briefly engage with his argument against the autonomy of technology, insofar as it is a step toward arguing for a radically open posthuman future. As opposed to the accelerationists and posthumanists, Roden's speculative posthumanism denies any conceptual or normative content to the historical trajectory of technological change. If Roden is correct, submitting to the machine promises nothing, means nothing. Any motivation to do so can only arise from an unsatisfiable "xenophilia," a desire for otherness.[4] But we shouldn't think that, *á la* Land, this desire for unthinkable otherness simply leaves desire unchecked to dissolve the rational human. Nor should we think that there could be a rational compulsion to become posthuman, leaving behind animality to transform thought, insofar as we can have no conviction that any such transformation would still count as thought. As Roden puts it, "Its scope is correspondingly indeterminate, perhaps closer to hand, and philosophically mute."[5]

Roden makes the case for the impossibility of simultaneously self-augmenting and autonomous technology by way of a critique of Jacques Ellul's robust technological determinism, which holds that technology is both. For Ellul, technology becomes self-augmenting by way of increasing complexity such that "no single agent or collective is able to exercise decisive control over the technical system," and yet the technical apparatuses we have are such that their functioning requires an increasing scaffolding of further artifacts and systems.[6] This increased complexity means that technology becomes progressively more *abstract*; new technical arrangements can be deployed in new contexts, for new functions, with new results. This abstraction is what obviates any sense of technical autonomy, that is, any concrete determinate "aims" of technology that might "control" human beings in the sense of dominating them in service to those ends. But it's worth noting that establishing the incompatibility of two features does not establish which fails to obtain. Perhaps technology is

---

[4] "Subtractive-Catastrophic Xenophilia," 46.
[5] "Subtractive-Catastrophic Xenophilia," 46.
[6] *Posthuman Life*, 156.

autonomous, embodying specific values, without being self-augmenting. This would be closer to Fisher's view. It is unclear whether a Promethean or accelerationist could accept it. Perhaps they must deny that humans can play a role in shaping, delaying, or—precisely—accelerating technological change. In any case, for the moment it seems that the option remains live.

A greater challenge for the Promethean in this regard is the very mode of intelligibility of technological change. Like biological explanation, it blurs the line between the manifest and scientific images. Technological artifacts, and their subsequent effects, can be explained simply in terms of what they *do*, or in terms of their *functions*, proprieties of use, intentions. So, perhaps, the categorial transformations that Christias or Brassier await will clarify how to conceive of the things we now see as technical artifacts and systems in utterly non-normative ways that render the question of technological determinism moot. For example, in some future ontology the systems that comprise the apparently teleologically structured activities and forms of life and technology may come to be seen as one single phenomenon, erasing the question of whether technology is "in control" of life, even removing questions of control entirely. Consider, for example, views of life on which what is central to life is its metabolic exchange with its environment; on such a view, life is essentially a machine and, in turn, a thermodynamic process. Or consider views of life that are essentially informational. Given the tight connection between thermodynamics and information we saw in Sect. 1.2, we can basically treat these in thermodynamic terms. In each case, while organisms may have mechanisms for resisting the entropic arrow of time through homeostasis, on the whole there is simply a probabilistic tendency toward equilibrium. Land himself seems to endorse such a view, claiming that "the history of thermodynamics is the history of technicizing commerce – of modernizing machines."[7] Causal laws are inessential to the picture, and so too are questions of "control." Nevertheless, despite morphological variations that arise in the process of resisting entropy, all technological change ultimately converges. I don't intend, here, to endorse either of these views, but simply to show that there are alternative approaches to technological "explanation" which might hew closer to Promethean and accelerationist views on which the historical trajectory of technology is not in our control but nevertheless is functionally "autonomous."

---

[7] "Machines and Technocultural Complexity: The Challenge of the Deleuze-Guattari Conjunction," 133.

Of course we have not arrived at any such radical categorial revision. And we need to make some sense of technological development in the meantime, both in order to better understand post-Landian posthumanism and to assess it. This is so even if the explanation must be inadequate. For Land, the nonlinear science of far-from-equilibrium, complex systems and their strange attractors might fit the bill. For other Deleuzians, evolutionary theory has seemed attractive, insofar as it also blurs the line between the manifest and scientific images. So, for example, in the highly influential *Deleuze and Philosophy*, both Howard Caygill and Keith Ansell-Pearson engage with evolution as an explanatory paradigm for understanding both life and technics, even if cautioning against simplistic over-identifications, in language that evokes precisely the sort of becoming-alien for which posthumanists strive.[8] So, for example:

> The notion of machinic evolution, therefore, does not refer specifically or exclusively to human contrivances, gadgets or tools, but rather to particular modes of evolution, such as symbiosis and contagion, and is not specific or peculiar to the human—machine relationship, since it also speaks of the machine—machine nexus and alterity. The 'machinic' is the mode of evolution that is specific and peculiar to the 'becoming' of alien life.[9]

It is not only posthumanists and Deleuzians, of course, who have attempted, at least metaphorically or analogically, to extend the theory of evolution to the process of technological development and change.[10] At its

---

[8] See Caygill, "The Topology of Selection: The Limits of Deleuze's Biophilosophy," and Ansell-Pearson, "Viroid Life: On Machines, Technics, and Evolution." I note these two, in particular, as they were both Land's interlocutors: Ansell-Pearson his colleague at Warwick and Caygill a fellow founding editor of the *Journal of Nietzsche Studies*.

[9] Ansell-Pearson, "Viroid Life," 193.

[10] See, e.g., George Basalla, *The Evolution of Technology* and Tessaleno C. Devezas, "Evolutionary Theory of Technological Change – State-of-the-art and new approaches." Interestingly, one of Deleuze's theoretical references (and thus, indirectly, one of Land's), Gilbert Simondon viewed the autonomization of technology *precisely* in evolutionary terms. However, his understanding was distinctly Bergsonian, and hence wildly heterodox (cf. Voss, "Gilbert Simondon and Different Senses of 'Evolution'"). We do not have space here to deal with the complications of such a view, especially because Simondon opposed his views to Wieners precisely because he, like Stiegler, resolutely defended *a priori* differences between the living and the technical object (*On the Mode of Existence of Technical Objects*, 51). It is worth noting that Andrea Bardin and Marco Ferrari have argued that Simondon actually articulates a philosophy of technology that allows us to resist the conception of social orga-

simplest, the basic idea is that, through processes of variation and natural selection, certain traits, forms, and features become predominant in a population. The *appearance* of purposiveness of biological traits does no explaining but is rather explained by the fact that these traits promote reproductive fitness.[11] If technological development is something like an evolutionary process, then innovation and deployment would in turn be something like the forces of chance— variation and drift—and the resulting systems and artifacts would be *selected for* in virtue of some further feature(s). On such a view, technological development would be just as "out of our control" as biological evolution, while still allowing for some degree of rational intervention; this is consistent with the process being at the same time highly path-dependent, leading to at least local convergence of traits: "runaway differentiation 'locks in' to an artificial destiny."[12] And just as the possibility space for biological variation shrinks as evolution is canalized, so too might the space of the technological future. Just like the heat death of the universe, even if technology's ends are nothing but a matter of overwhelming odds, we can still bet on them.

So, consider that much of Roden's argument against technological autonomy hinges on the fact that Ellul takes the aim of technology—or "technique," in Ellul's expansive sense—to be *efficiency*. Yet, by Roden's light, technology becomes self-augmenting in virtue of its increasing complexity and abstraction, which makes any substantial goal of "efficiency" on technology's part impossible; abstract techniques are "multistable" and can be put toward any number of ends. On an evolutionary view, however, this makes perfect sense; "efficiency" is a higher-order goal in the same way that

---

nization forced upon us by a cybernetic project very much like the one I will articulate in the following subsections (cf. "Governing Progress"). I am not sanguine about its prospects.

[11] I don't believe that any of this implies any position with respect to the ultimate units of selection, whether genes or organisms or what have you, nor on issues of "genetic determinism." Indeed, everything I say here is fully consistent with the construction of evolutionary niches, massive epigenetic contributions to development, etc.

[12] Land, "Machines and Technocultural Complexity," 133. On "path dependence" in technological development, see Rycroft and Kash, "Path Dependence in the Innovation of Complex Technologies," and Araujo and Harrison, "Path Dependence, Agency and Technological Development." At its most basic, the idea resembles Hughes' concept of "technological momentum," wherein buy-in to and deployment of large-scale technologies can have structuring effects on society, which in turn constrain or canalize technological development. See Hughes, "Technological Momentum." Rycroft and Kash's account seems especially resonant with the Landian emphasis on complex systems, noting that small changes can have outsized influence on developmental paths, facilitated by feedback loops.

"reproductive fitness" is. They don't indicate concrete *teloi* but rather are *selected for*. Many different features can make an organism reproductively fit, whatever chance path of development by which they emerge. In the same way, many different features might make an artifact or technical system efficient. The technical case is more complex, as there are many different ways in which we might understand efficiency but, bracketing that for a moment, we might nevertheless think that artifacts and systems can be "selected for" their efficiency. So, for example, a system that transmits information more efficiently than any other, in a larger context in which transmitting information is an important background condition, would be selected for. Some, like Allan Dafoe, suggest more determinate conditions, such "military-economic selection," suggesting that the kind of efficiency that matters is efficiency in achieving military and economic ends.[13] At any rate, if something like this is true, one can see how technology might, then, force human activity toward its own higher-order ends of efficiency, even if through processes of random chance and selection, if the costs of opting out of more complex systems are unreasonable.

Once again, I am not claiming that accelerationists accept such a view. Negarestani, for example, might resist transposing his dialectic of intelligence into the messy realm of contingent evolution. The point is simply that *if* such a way of thinking about the way that technology develops and recruits agents toward its own ends is available at all, then it is available to the Promethean, and it is at least possible to hold that technology is not "in control" but nevertheless might drive human agency in a determinate direction. And at the very least, Private like Wolfendale precisely describe the technological scaffolding and extension of intelligence in evolutionary language, as a matter of "adaptation." And it is precisely in terms of the evolution of our information technologies that we shall view the transition from neoliberal to surveillance capitalism in the history of intelligence.

5.2 The Machinic Revolution of Neoliberal Cybernetics

However Prometheans and accelerationists make sense of the dynamics underlying the historical trajectory, they clearly privilege a subset of technologies in their thought. Indeed, these could be called "technologies of

[13] Allan Dafoe, "On Technological Determinism: A Typology, Scope, Conditions, and a Mechanism."

thought," if the very concept of thought was not in question. Here they once again follow Lyotard, who not only rehabilitated capitalism as a multiplier and intensifier of *desire* but also placed information technology, the sciences of complexity, and "the perfection of intelligent terminals" at the heart of postmodern posthumanism.[14] We've seen already the persistent fascination with artificial intelligence and antihumanist "intelligenic" processes. At stake is whether this intelligence is an *expression of thought*, in the classical or folk-psychological sense of the term, explicated either in phenomenological terms or in the language of "propositional attitudes," or if it is rather a matter of comportment or behavior, perhaps facilitated by thought in some circumstances but also held back, something for which thought is at best optional, but essentially epiphenomenal. The future of post-Landian posthumanism hinges on the future of such technologies, and the manner in which they conscript us—in some sense of "us" —to their operations.

Obviously, computers as artifacts, along with the software and programs that allow them to operate, including contemporary machine learning algorithms and narrow artificial intelligence, are the technical objects and systems that explicitly concern Prometheans and accelerationists. And the history of AI can be viewed as a competition between two competing paradigms, the "symbolic" and the "connectionist." The former is modeled on formal logic, with programs following decision trees in response to symbolic inputs. While of course highly abstracted, these might reasonably be taken as an attempt to instantiate a classical image of discursive or propositional thought, with its "fundamental assumption that language-like structures of some kind constitute the basic or most important form of representation... and the correlative assumption that cognition consists in the manipulation of those representations by structure-sensitive rules."[15] This is, more or less, the manifest image of the space of reasons. On the other hand, there is the "connectionist" approach to AI, explicitly cybernetic in its origins, based on the idea that computer programs could function like the brain, from which "neural networks" derive. For such systems, while there is a logical structure, what is crucial is not that structure but the threshold of activation of "neurons," which can be reinforced by repeated activation. It was precisely this feedback-driven cybernetic vision of AI that fed Land's feverish theoretical imagination:

---

[14] *The Postmodern Condition: A Report on Knowledge*, 4.
[15] Paul M. Churchland, "On the Nature of Theories: A Neurocomputational Perspective," 252.

Positive feedback is the elementary diagram for self regenerating circuitry, cumulative interaction, autocatalysis, self-reinforcing processes, escalation, schismogenesis, self-organization, compressive series, deuterolearning, chain-reaction, vicious circles, and cybergenics. Such processes resist historical intelligibility, since they obsolesce every possible analogue for anticipated change. The future of runaway processes derides all precedent, even when deploying it as camouflage.[16]

What we shall witness is the realization of the cybernetic imagination in the project of neoliberalism.

The history of AI, as is well known, is often described as a succession of "summers" and "winters," the former indicating periods of substantial state and/or industry funding, usually accompanied by at least some substantial growth in AI capabilities. So, for example, early and promising developments in "expert systems," a form of symbolic AI supplemented with large banks of stored information, led in part to a surge in funding from DARPA's Strategic Computing Initiative from 1983 to 1993, and disappointment with diminishing performance returns led to an AI "winter" for nearly two decades. Understandably, attitudes cooled considerably toward symbolic AI, which to some still seemed like the only game in town.[17]

However, a lack of funding could not bring AI research to a halt. Connectionist approaches to artificial neural networks received a great deal of attention through the 2000s, under the auspices of "informatics" and "analytics," to avoid negative connotations associated with overblown AI hype.[18] Indeed, the Big Data Revolution was, in effect, an AI revolution. If one of the early limitations of neural networks was the necessity of massive amounts of training data, the massive growth in surveillance and digital storage of the 1990s and 2000s would unleash their potential.

---

[16] "Machinic Desire," 330. Land is perhaps overstating things here; it is not at all clear that these processes resist historical intelligibility. Negative feedback loops keep systems within limits, and not all positive feedback processes are "runaway." Moreover, while far-from-equilibrium systems can—occasionally by virtue of positive feedback loops—undergo "phase shifts," these are all still perfectly intelligible. Nonlinearity is not incomprehensibility.

[17] So, for example, in her seminal 1984 study of the ways in which computing technologies impact our self-understanding, Sherry Turkle is utterly dismissive of the sort of connectionist "parallel processing" that informed the self-image of some of her subjects as "only a dream" (*The Second Self: Computers and the Human Spirit*, 259). This is not to critique Turkle, whose work in connecting AI and psychoanalysis has informed my own thinking in this book.

[18] Toosi et al., "A Brief History of AI: How to Prevent Another Winter – A Critical review."

Since the early 2010s we have witnessed the "revenge of the neurons" as machine learning and, especially, deep learning algorithms have become increasingly capable and, more importantly, widely deployed, from recommender systems to data-driven science to autonomous vehicles.[19] While AI development has had its waves and cycles, and things may change, it seems that Land's vision has won out:

> Where formalist AI is incremental and progressive, caged in the pre-specified data-bases and processing routines of expert systems, connectionist or antiformalist AI is explosive and opportunistic: engineering time. It breaks out non-locally across intelligent networks that are technical but no longer technological, since they elude both theory-dependency and behavioural predictability. No one knows what to expect. The Turing-cops have to model net-sentience irruption as ultimate nuclear accident: core meltdown, loss of control, soft-autoreplication feeding regeneratively into social fission, trashed meat all over the place.[20]

In particular, deep learning algorithms, which are capable of unsupervised and self-supervised learning, in some cases without relying on the work of human "labelers" to categorize training data, are examples of the forms of self-organizing, complex processes and systems that fascinate accelerationists. They were of course not alone in this. As William Connolly puts it, the "compass of neoliberalism itself includes all those who celebrate the self-organizing powers and impersonal rationality of markets while 'limiting' state and other political involvements to shoring up market processes."[21]

Unsurprisingly, Hayek was deeply inspired by cybernetics, which he in turn traced back to eighteenth-century political economy:

> This mutual adjustment of individual plans is brought about by what, since the physical sciences have also begun to concern themselves with spontane-

---

[19] Dominique Cardon, Jean-Philippe Cointet, and Antoine Mazières, "La revanche des neurones. L'invention des machines inductives et la controverse de l'intelligence artificielle." See also Terrence Sejnowski, *The Deep Learning Revolution*. I don't distinguish here between the learning algorithms that we would describe as "data analytics" and the large-language models that have co-opted the term "AI." Though the transformer architectures that distinguish the latter are impressive achievements, they do not constitute a change in paradigm. See Sejnowski's *ChatGPT and the Future of Language*, especially Ch. 6.
[20] "Meltdown" 450.
[21] *The Fragility of Things*, 52.

ous orders, or 'self-organizing systems', we have learnt to call 'negative feedback.' Indeed, as intelligent biologists acknowledge, "long before" Claude Bernard, Clerk Maxwell, Walter B. Cannon, or Norbert Wiener developed cybernetics, "Adam Smith had just as clearly used the idea... The 'invisible hand' that regulates prices to a nicety is clearly this idea. In a free market, says Smith in effect, prices are regulated by negative feedback."[22]

Liberalism (and neoliberalism) and cybernetics are thoroughly entwined. And just as liberalism and cybernetics were intertwined, so too were neoliberalism and the development of connectionism. It's worth remembering that while contemporary discussion of "black box" neural networks and machine learning algorithms tends to focus on their performance, in industry and science, in our devices and for our convenience, the discipline of AI lies at the origins of cognitive as well as computer science. And early cybernetic ideas about connectionism were just as much about understanding the structure of the brain as they were engaged in reproducing its behavior in machines. As Matteo Pasquinelli puts it, with respect to Hayek's *Sensory Order*, "a neoliberal economist... provided the most systematic treatise on connectionism or, as it would later be known, the paradigm of artificial neural networks."[23]

Nevertheless, Pasquinelli goes on to claim that "Hayek *stole* pattern recognition and transformed it into a neoliberal principle of market regulation."[24] This, I will suggest, is mistaken, but it is part of Pasquinelli's larger critique of the imbrication of AI and contemporary capitalism. Like Land, Pasquinelli and other critics such as Nick Dyer-Witheford, Atle Mikkola Kjøsen, and James Steinhoff take it that AI is simply a force multiplier for capital.[25] Unlike Land, these critics share the same general approach, which is to view AI technologies as instruments that aim at the

---

[22] "Competition as a Discovery Procedure," 309. This lecture was given in 1968. It's worth noting that the next year later, Otto Mayr independently likewise traced the innovation and deployment of feedback-controlled technologies to social orders in which "liberal" ideas of ordered individual freedom were prominent. See *The Origins of Feedback Control*. In that text, he was somewhat tentative, but his 1986 *Authority, Liberty, and Automatic Machinery in Early Modern Europe* would much more confidently make the case.
[23] *The Eye of the Master: A Social History of Artificial Intelligence*, 183.
[24] *The Eye of the Master*, 183, emphasis mine.
[25] See Dyer-Witheford, Kjøsen, and Steinhoff, *Inhuman Power: Artificial Intelligence and the Future of Capitalism*, and Steinhoff, *Automation and Autonomy: Labour, Capital and Machines in the Artificial Intelligence Industry*. While both of these, along with Pasquinelli, view AI as a form of automation, the latter is particularly compelling in discussing AI as an industry unto itself.

automation of labor; their target is not just the increasing disempowerment of labor but those who think that this automation can be harnessed for the utopian dream of a post-work world. Against Prometheanism, they all seem quite skeptical of the emancipatory potential of AI. But the point is that they take AI, in automating the production of knowledge, to be like any other functional technology in that it can replace human labor. But, I think, this is to misunderstand the fundamental issue. "Automating science" and all sorts of other human activities, economic and non-economic, is a dimension of the rise of AI, and certainly a concern of the founders of cybernetics. But, as Weiner says, while "automatization is one phase of" cybernetics, it is still only one phase.[26]

More strictly, neoliberals like Hayek were interested in the market because they were interested in *market orders*, that is, self-organizing social systems for which the mechanism coordinating the behavior of self-interested, severely cognitively limited agents is the competitive market. This is why, as we saw above, Fisher could castigate contemporary capitalism as an *anti-market*, a structure that hinders the process of self-organization. It is not that Hayek "stole pattern recognition and transformed it into a principle of neoliberal market recognition" but rather that he grasped after principles of self-regulation by which to *organize behavior* and found them in cybernetics.

The Sensory Order, in fact, is not a work on the market, or on computing, but the human mind, even if inspired by cyberneticians like Warren McCulloch and Walter Pitts, who produced the first computational model of neuronal activity. The point is that, as we saw in Sect. 3.2, Hayek was particularly entranced by the nascent idea of neural nets that could respond intelligently to inputs (data) without a necessary detour through anything corresponding to our manifest image of thoughts connected by inferential relations. Frank Rosenblatt, in turn, actually produced the Perceptron, the first artificial neural network, and was himself inspired by Hayek but also by Donald Hebb's classic work, *The Organization of Behavior*.

While best known for introducing Hebbian learning, it's important to remember the aims and background of Hebb's investigation of the brain. To begin, it was thoroughly behaviorist and mechanistic. Hebb attributed to psychology "the task of understanding behavior and reducing the vagaries of human thought to a mechanical process of cause and effect."[27] For

---

[26] Gordon S. Brown and Norbert Wiener, "Automation, 1955: A Retrospective," 383.
[27] *The Organization of Behavior*, xi.

example, Hebb is critical of Freud but only because the concepts of Id, Ego, and Superego might lead theorists to think that intentionality is a fundamental feature of mind.[28] Rather, "Modern psychology takes completely for granted that behavior and neural function are perfectly correlated, that one is completely caused by the other.... 'Mind' can only be regarded, for scientific purposes, as the activity of the brain."[29] In classic behaviorist fashion, even *desire*—for food, or sex—is reduced to behavioral dispositions of "responsiveness to... stimulation," in turn awaiting their explication in neuronal terms.[30] Hebb is explicit that the aim of the project of psychology is to reduce behavior to the control of the brain.[31] In brief, the aim is to reduce intentional action to behavior, and in turn to explain the production and regulation of that behavior in terms of the activation of neural networks.

While Rosenblatt's specific debt to Hayek and Hebb is more technical, the point is that the entire connectionist project is aimed at coordinating behavior through external stimuli, without the mediation of knowledge or conscious control. Markets are one means of doing so, using the external stimuli of consumer and producer behavior to, in turn, produce further stimuli—whether conceived as signs, information, or simply "prices"— allowing individuals to regulate themselves. So if Hayek thinks of the market as something like a computer program, such a primitive version of AI is necessary *because* humans are incapable of effectively or efficiently organizing their behavior themselves; what information they have is distributed across a social body that no one has access to, and even that "information" is not, generally, propositional knowledge. What knowledge we have is largely a matter of "know-how," that is, "mental factors which govern all our acting and thinking without being known to us, and which can be described only as abstract rules guiding us without our knowledge."[32] That is to say, the aim is producing a social order, one in which the manifest image of thought is no longer necessary. Automation plays a part in that, but more broadly, neoliberalism, as an epistemic project, is part and parcel of the cybernetic project.[33]

---

[28] *The Organization of Behavior*, xii–xiii.
[29] *The Organization of Behavior*, xiii–xiv.
[30] *The Organization of Behavior*, 207.
[31] *The Organization of Behavior*, xi.
[32] Hayek, "The Primacy of the Abstract," 318.
[33] One might call this "cybernetic society." Michael Thompson has made critical use of the term to describe an anti-individualist, thoroughly integrated, hierarchically and bureaucratically controlled social order (see *The Twilight of the Self*, 35), but this, I think, misconstrues the *mode* of cybernetic control, in which individuals are integrated through, precisely, individual self-regulation and self-organization.

If humans are connectionist machines, they require another connectionist machine to coordinate them. Society needs a mind, but it cannot be human. For both Hayek and Land, the market is the "immanent intelligence" of the technological system we find ourselves in:

> Markets are part of the infrastructure - its immanent intelligence – and thus entirely indissociable from the forces of production. It makes no more sense to try to rescue the economy from capital by demarketization than it does to liberate the proletarian from false consciousness by decortication. In neither case would one be left with anything except a radically dysfunctional wreck, terminally shut-down hardware. Machinic revolution must therefore go in the opposite direction to socialistic regulation; pressing towards ever more uninhibited marketization of the processes that are tearing down the social field....[34]

Here Land echoes Hayek, if filtered through cyberpunk and French antihumanism:

> It was men's submission to the impersonal forces of the market that in the past has made possible the growth of a civilization which without this could not have developed; it is by thus submitting that we are every day helping to build something that is greater than anyone of us can fully comprehend.[35]

Insofar as neoliberalism has been the dominant political form of the modern Western world, we are caught in the "machinic revolution" of the market. While the previous section argued that we can legitimately view the development of technology as "autonomous," the aim of this section has been to sketch a view of the twin histories of AI and neoliberalism as part of the longer history of the becoming-cybernetic, or becoming-self-organized-through-self-regulation, of the self and the social order to which it belongs. So, Land is *partially* right.

In the following section, I will consider the Promethean challenge to accelerationism, pitched in terms of technological *alternatives* to the market. We shall see how these are already captured in the webs of surveillance capitalism, and continue the cybernetic project of producing mindless order.

---

[34] Land, "Machinic Desire," 340.
[35] *The Road to Serfdom*, 210.

5.3 MONT PELERIN MULTIVAC: THE CALCULATION PROBLEM AND THE STYGIAN INTELLIGENCE OF PLATFORM CAPITALISM

Gabriel Oliva, in their important review of Hayek's relationship to the idea of feedback-control, distinguishes between cybernetics proper and the "general systems theory" of Ludwig von Bertalanffy. The distinction, which is supposed to help explain Hayek's account of the evolution of the social order, is not in itself important for us here; systems theory, complex adaptive systems, autopoiesis, second-order cybernetics, etc., are all importantly different but can still be understood as extensions, transformations, or continuations of the universal ambitions of the cybernetic project. Oliva's point is that, while cybernetics focuses on self-regulation in terms of *homeostasis*, the maintenance of an internal order through feedback mechanisms, general systems theory introduced "heterostasis"— or, as it is more commonly called, allostasis—as the *modification* of that order in relation to external change. Structures may radically transform, in ways that Bertalanffy thought transcended negative feedback. Here we see the "positive feedback," the critical phase shifts, that so enchanted Land with their revolutionary potential. More specifically, Bertalanffy described "open systems" in thermodynamic terms, modifying themselves by feeding on "negative entropy"—or information—in the environment, "importing complex organic molecules, using their energy, and rendering back the simpler end products to the environment."[36] Obviously, the extension of this language of metabolic and informational exchange is metaphorical, and Hayek makes use of broadly similar concepts to explain the evolutionary emergence of the market. We are still trying to answer Asimov's last question.

In this section, I'll be suggesting that these entropic metaphors still capture Promethean attempts to think beyond the market, and the transformations of this "open system" into what has been called, variously, platform or surveillance capitalism. Against both accelerationists and Prometheans, the capture of behavior by contemporary AI in its deployment as a tool to *know* human beings, to produce the truth about human beings, does not promise us escape.

---

[36] "The Road to Servo-Mechanism: The Influence of Cybernetics on Hayek from The Sensory Order to the Social Order," 32.

As we have seen, accelerationists and Prometheans believe that the revolutionary dynamics of technology faced obstacles. For Land, it is "government," or politics more generally.[37] Land (and Fisher, maybe) seems to think that we could unleash the market, or a similar sort of truth-machine, from its various obstacles, to catalyze the technological production of the future. In this, he was very much following Hayek's lead. Part of what makes Hayek an unorthodox economist is not simply his hostility to scientific and quantitative formalism, but also his underlying focus on economic action as a species of human conduct. The interest in market orders is an interest in "catallaxy," a fundamental form of action "reconciling" individuals with opposed goals.[38] This is an inheritance from the praxeology of Ludwig von Mises; besides his framing of economics within the broader question of human action, von Mises' great contribution to the history of thought is his intervention in the so-called socialist calculation debate. Von Mises was the first to point out that socialist governments had no possible way to allocate resources efficiently, in ways that wouldn't require repression; markets, and the private property relations that sustain them, were the *only* way in which to do so. Hayek's greatest achievement, arguably, was to construe this as a matter of knowledge, and to envision the market as an algorithm for organizing that knowledge and producing prices allowing for coordinated behavior. Further, contra market socialists like Oskar Lange, Hayek points out that no centralized planner could possibly collect the relevant data or perform the calculations required to produce these prices.

Pasquinelli is correct to point out the origins of Hayek's connectionism in this debate.[39] However, he situates the debate in a longer Marxist critique of political economy, in which the automation of thought in neural networks constitutes a stage in the process of "real abstraction," as a reflection of the social division of labor.[40] The information age, on his view, represents the automation—and hence the intensified exploitation—of the "general intellect" produced by the abstract labor required by the Industrial Revolution.[41] In this, he adopts a fairly classical Marxism, if closely aligned to the "operaismo" and autonomist traditions of Marxist

---

[37] "Government is isomorphic with top-down AI and increasingly scrambled with it" ("Meltdown," 352).
[38] "Competition as a Discovery Procedure," 307.
[39] *The Eye of the Master*, 199–201.
[40] *The Eye of the Master*, 204.
[41] *The Eye of the Master*, 245.

and post-Marxist thought with their focus on the potential of know-how and "immaterial labor" in an information economy organized around cognitive and behavioral tasks. In what follows, we largely avoid the question of labor, insofar as we will adopt a broadly Foucauldian lens; what concerns us is not the specifically economic dimensions of the problem of knowledge, but the connections between social order and subjectivity that make possible any sort of conduct.

Like Land, Prometheans also believe that there are obstacles holding back the power of technology, and similarly trace their project back to the socialist calculation debate. Unlike for Land, it is not government that is the obstacle, but technical limitations. So, for example, Srnicek and Williams, like good socialists, find inspiration in cybernetic attempts to solve the calculation problem. They continue to embrace the cybernetic imaginary:

> The Chilean Project Cybersyn is emblematic of this experimental attitude- fusing advanced cybernetic technologies, with sophisticated economic modelling, and a democratic plat form instantiated in the technological infrastructure itself. Similar experiments were conducted in 1950s-60s Soviet economics as well, employing cybernetics and linear programming in an attempt to overcome the new problems faced by the first communist economy. That both of these were ultimately unsuccessful can be traced to the political and technological constraints these early cyberneticians operated under.[42]

Cybersyn was an attempt by Allende's socialist government in Chile to solve the calculation problem—the problem that Hayek thought that the market would solve, calculating prices so as to efficiently satisfy the desires of citizen-consumers—by way of computation and cybernetics; if humans are epistemically limited, then in principle the computer can serve in their stead, assessing their demand to calculate supply. This was by no means an aberration. Oskar Lange, a key contributor to the debate about the calculation problem, was explicit that the (cybernetic, connectionist) computer could replace the market:

---

[42] "#Accelerate: Manifesto for an Accelerationist Politics," 357. Williams remains enthralled by the strategy of the Mont Pelerin collective more broadly; cf. "Strategy without a Strategiser," 22–23.

The market process... appears old-fashioned. Indeed, it may be considered as a computing device of the pre-electronic age. The market mechanism... really played the role of a computing device for solving a system of simultaneous equations... based on a feedback principle operating so as to gradually eliminate deviations from equilibrium. It was envisaged that the process would operate like a servo-mechanism, which, through feedback action, automatically eliminates disturbances.[43]

The market was a computer, an artificial intelligence. Socialists thought that the cybernetic computer could replace it, allowing for the sort of planning that would make possible a genuine economic democracy. They were half right.

At any rate, for Land and for the Prometheans, artificial intelligence succeeds Multivac in the posthumanist cybernetic imaginary. Can entropy be reduced? Can the social be (self-)organized? Indeed, Cybersyn was in part the project of the prominent cyberneticist Stafford Beer, for whom cybernetics captured not only the self-regulation of organisms but also of corporate firms.[44] In Chile this socialist project was destroyed by a coup, but the cybernetic project was continued by the "Chicago Boys" who helped install an early neoliberal economy. While Friedman and the Chicago school represented a distinct approach to cybernetics from Hayek, the latter still visited Chile during the Pinochet regime.[45]

In his lectures on the origins of neoliberalism, Foucault—controversially—claims that, while governmentality in the twentieth century is liberal, and eventually neoliberal, there is no competing *socialist* governmentality.[46] Commentators have wondered what he meant by this. But I think this is an example; for both Hayek, the neoliberal, and Lange, the socialist, the question is how to coordinate social behavior, and the

---

[43] "The Computer and the Market," 158–159.
[44] *The Brain of the Firm*, 75–76. See also Eden Medina, *Cybernetic Revolutionaries: Technology and Politics in Allende's Chile*, and Max Hancock, "Spontaneity and Control: Friedrich Hayek, Stafford Beer, and the Principles of Self-Organization." It is also worth noting that Beer is credited with the "cybernetic dictum" that "the purpose of a system is what it does," an encapsulation of the ways in which the cybernetic project has no place for normativity. He does so in a text that not only relays his experience with Allende, but relates the Chilean coup of September 11, 1973 to the attacks of September 11, 2001, explaining the latter as a form of feedback response to what the US system does: "this amplification of system turned hijacked planes into guided missiles." See "What is Cybernetics?"
[45] Cf. Bruce Caldwell and Leonidas Montes, "Friedrich Hayek and his visits to Chile."
[46] Foucault, *The Birth of Biopolitics*, 92–94.

answer is to efficiently fulfill their desires. For all its rationalism, Prometheanism seems to fall back into the desire-centric trap of accelerationism. To govern by desire, to make *interest* the principle of self-regulation of individuals, and to have markets respond intelligently to those desires: this is the neoliberal, cybernetic project. Indeed, Foucault coins the term "governmentality" to describe "the conduct of conduct," that is, precisely to name the sort of bottom-up, feedback-driven form of self-regulating power he thought was embodied in the neoliberal market project. If that description evokes cybernetics, it should: he, like Wiener before him, draws his nomenclature precisely from the Greek *kubernetes*, the steersman or captain.[47]

If the market functions as Multivac in the cybernetic imagination, collecting ever more data in its attempts to fend off entropy and maintain some sort of organization, we should remember that in Asimov's story, Multivac is succeeded by a number of further evolved machines, from Microvac to Galactic AC to Cosmic AC to, simply, "AC." And just as Hayek thought the market was a fortuitous result of unplanned adaptation, a product of cultural evolution, so that evolution might continue into *new* algorithms and technologies. And, indeed, not only has our technological situation altered but so too have the aims and ends of the market, and our understandings of the problem that the market was supposed to solve. It is now, I think, accepted that Hayek and other neoliberals believed that the market, as important as it is for coordinating behavior, was not and could not be the *only* source of behavioral regulation. Individual market behavior takes places against a broader cultural background of norms. For Hayek himself, the market order, being itself unplanned, also required the evolution of a "market morality."[48] Melinda Cooper has shown that the rise of neoliberal policies required the cultivation of a traditional, individual, and family-centric *ethos*.[49] More radically, the libertarian nationalist Peter Brimelow referred to this broader context as the "metamarket," and attempted to "extend Hayek's argument," that a racially homogeneous metamarket would be most conducive to economic performance.[50] While we will not be focusing on the cultural or

---

[47] Foucault, *Security, Territory, Population*, 122–123; Wiener, *Cybernetics, or Control and Communication in the Animal and the Machine*, 11.
[48] Naomi Beck, "Be Fruitful and Multiply: Growth, Reason, and Cultural Group Selection in Hayek and Darwin," 415.
[49] Cf. *Family Values: Between Neoliberalism and the New Social Conservatism*.
[50] Slobodian, "The Ethno-economy," 16.

moral dimensions of the metamarket, nevertheless, to return to the cybernetic metaphor, the market has been subjected to allostatic pressure, and the production of a homeostatic social order requires a change in the "rules of the game."

We are seeing this change play out in the rise of surveillance, or platform, capitalism. Just as some elements of our contemporary technosocial conjuncture validate Land, so some validate the Prometheans. AI, the machine, the computer, *has* replaced the market, or is in the process of doing so. More specifically, the *platform* has replaced the computer. Fisher was right that contemporary capitalism is no longer fundamentally about markets. The accelerationists were wrong. Fortunately for them and unfortunately for Fisher, this doesn't matter.

Contemporary capitalism is no longer industrial, or simply neoliberal. It goes by many names, like "data capitalism," "digital capitalism," "platform capitalism," or "surveillance capitalism." Shoshanna Zuboff uses the latter. I think her analyses are particularly insightful, even if the terminology overemphasizes "surveillance" and obscures their deeper importance, so I will use the term "platform capitalism" instead.[51] Four important changes matter for platform capitalism: the explosion of sensor technology and Big Data, the rise of machine learning algorithms for data analytics, the development of actuator technology along with sensor technology, and the integration of the behavioral sciences into the economics of human action.

Over the last three decades, the dominant form of information technology has become not merely the computer, or the Internet, but the *platform*, the virtual space and its infrastructure through which we interact online. These platforms not only collect data from all sorts of external sensor technologies, but also through our interactions with them. In this way, they function like neoliberal markets; where ordoliberals and the Mont Pelerin neoliberals thought we needed to expand competitive markets to all of society, contemporary platform capitalism has already succeeded. We interact with them all the time; we don't need to disclose our desires through market interactions because we are constantly disclosing our behavior all the time. "The price-system-whose epistemological

---

[51] To his credit, Srnicek attempted an early analysis of platform capitalism, and emphasizes the key point that they are founded on new extractive industry, namely, data extraction. Unfortunately, he fails to delve deeper into the dynamics and aims of this process in the ways that Zuboff and others do. Cf. *Platform Capitalism*.

function has long been understood-thus transitions into reflexively self-enhancing technological hyper-cognition."[52] As all reality becomes a platform, capitalism builds its own Cosmic AC.

Perhaps the price system "transitions into reflexively self-enhancing technological hyper-cognition" on its own, or perhaps its "well-known" epistemological function has been rethought. At the very least, the technological limitations in virtue of which the market, qua price system, has to perform its function might need to be rethought. Where the point of the price system was to deal efficiently and effectively with the deep ignorance of human agents and the fact that what knowledge they possess is widely distributed among the population, platform capitalism, exploiting the Big Data revolution and rise of machine learning, aims to "challenge the classic quid pro quo of freedom for ignorance."

> When it comes to surveillance capitalist operations, the 'market' is no longer invisible, certainly not in the way that Smith or Hayek imagined. The competitive struggle among surveillance capitalists produces the compulsion toward totality. Total information tends toward certainty and the promise of guaranteed outcomes. These operations mean that the supply and demand of behavioral futures markets are rendered in infinite detail. Surveillance capitalism thus replaces mystery with certainty as it substitutes rendition, behavioral modification, and prediction for the old 'unsurveyable pattern.' This is a fundamental reversal of the classic ideal of the 'market' as intrinsically unknowable.[53]

It is striking how this dimension of platform capitalism has been overlooked by posthumanists. Lange's appeal to computers failed to convince many at the time, but over the past several years, the increasing capacities of AI have reopened the calculation debate, long thought to have been won by the Austrians. The topic remains contentious, of course, but the debate remains focused on the plausibility of something like cyber-socialism or cyber-communism.[54] Yet the project of Cybersyn is coming to fruition, not in the hands of a socialist government supported by a

[52] Land, "Teleoplexy: Notes on Acceleration," 516.
[53] Zuboff, *The Age of Surveillance Capitalism*, 497.
[54] See, for example, Brennan-Marquez and Susser, "Privacy, Autonomy, and the Dissolution of the Market," Acemoglu, "Hayek and Artificial Intelligence," Boettke and Candela, "The Feasibility of Technosocialism," Lambert and Fegley, "Economic Calculation in the Light of Advances in Big Data and Artificial Intelligence," and Dapprich and Greenwood, "Cybersocialism and the Future of the Socialist Calculation Debate."

5 HOW TO DELETE THE MANIFEST IMAGE: A POLITICAL ECONOMY... 109

collectivist ethos, but rather in those of the entrepreneurs of Silicon Valley, cultivating an ethos of mindless engagement.

More than 35 years ago, Zuboff had already warned us about the rise of what she called "informating," the process by which the technologies that served to automate functions in the modern firm, being integrated with more novel information and communication technologies, began to collect surplus data, which could then feed backwards into the firm to improve internal performance, in a process she calls "informating." This seemed to happen of its own accord. Managers

> believed that the technology would unleash an irreversible trend... In the absence of a management strategy to exploit this new potential, they maintained their enthusiasm for the autonomous power of an informating technology.[55]

This new informating technology rendered our behavior legible in new ways, and even decades ago Zuboff worried that "the managerial need for certainty will colonize the new behavioral text and finally convert it into another opportunity for enhanced control."[56] She would eventually learn how well founded that worry was.

Indeed, it is almost as if there was a *technological*, or *techno-rational*, imperative at work in the collection of all this data. Marion Fourcade and Kieran Healy note that, with the rise of information and communication technology, *even without knowing why or what they were collecting data for*, organizations largely conformed to the "data imperative," relentlessly extracting and stockpiling data that would serve as a resource for predictive analytics.[57] More tellingly, they describe the collection of data, and the processes of valuation, coordination, and subsequent value-extraction, as "seeing like market." If, in the larger neoliberal epistemic project, the role of the market was originally to collect data, value, and coordinate activity, those roles have been taken over by our post-market technologies. And, given that the cybernetic approach to coordinating activity works by means of prediction and feedback, the drive to collect ever more data to refine those predictions has been a part of it since the start.[58]

---

[55] Zuboff, *In the Age of the Smart Machine*, 385.
[56] *In the Age of the Smart Machine*, 386.
[57] Marion Fourcade and Kieran Healy, "Seeing like a market," 16.
[58] See Galison, "The Ontology of the Enemy." The very first stirrings of the "cybernetic vision" could be felt in Norbert Wiener's attempt to construct feedback-driven anti-aircraft

And, of course, a use for all this data would be found. By offering us free access—for social media, countless apps, etc.—platform capitalism "renders" our experience into "behavioral data," and—more specifically— into "behavioral surplus."[59] But mere behavioral data is not the endgame. The rapid growth in knowledge in the behavioral sciences allows tech companies to make *predictions* based on our behavior: a predictive power vastly enhanced by contemporary AI.[60] What is produced, then, are "markets of behavioral prediction."[61] However, this is not the endgame either. Once we are alienated from our experience in this way, that is, once the raw material of the manifest image becomes commodified as behavioral data, platform capitalism sets itself greater goals: "not the automation of society, as some might think, but rather the replacement of society with machine action dictated by economic imperatives."[62] Once again, the collaboration of data analytics and behavioral science makes this possible; in order to improve the accuracy of its predictions, platform capitalism doesn't only refine its tools but refines reality, using various nudging and herding techniques to *shape* our behavior, beneath our rational control, bringing it closer to predictions, and in turn more susceptible to further fine-tuning:

> The competitive dynamics of this new order require economies of action that operate to configure human behavior in ways that facilitate predictability. These operations grow more muscular with the escalation of competitive intensity, driving the evolution of predictability toward certainty. They are made manifest in a ubiquitous digital architecture of behavior modification owned and operated by surveillance capital… indecipherable, and largely hidden.[63]

Here one might object that we have moved far away from the neoliberal, cybernetic project, which aimed at providing the conditions for social

---

weapons that would themselves model enemy intentions with feedback loops. These were, ultimately, a failure, but the diagnosis of that failure lay in a lack of sufficient data.

[59] Zuboff, "Surveillance Capitalism and the Challenge of Collective Action," 3.

[60] Indeed, contemporary AI algorithms are essentially "prediction engines." Cf. Ajay Agrawal, Joshua Gans, and Avi Goldfarb, *Prediction Engines: The Simple Economics of Artificial Intelligence*.

[61] Zuboff, "Big Other: Surveillance Capitalism and the Prospects of an Informational Civilization," 85.

[62] Zuboff, *The Age of Surveillance Capitalism*, 220.

[63] Zuboff, "'We Make Them Dance': Surveillance Capitalism, the Rise of Instrumentarian Power, and the Threat to Human Rights," 38.

self-organization by allowing maximum freedom. Aren't we far removed from *markets*? And "nudging" is described by the scholars who coined the term as "libertarian paternalism."[64] But wouldn't any sort of paternalism, "libertarian" or not, be a serious divergence from neoliberalism? I don't think so. The point of the cybernetic project is the cultivation of *self-organization as a mode of control*. The conditions under which this self-organization takes place have changed and, thus, so too have the modes of this self-organization, but they remain part of a coherent trajectory. So, for example, despite winning a Nobel Prize, Hayek and the Austrian school have always been marginal in mainstream economics. And the calculation problem that so worried them was a function of ignorance, distributed knowledge, and complexity. As noted, some have recognized the computational powers of AI as a breakthrough, but in many ways the importance of platform capitalism and the surveillance society in which it took root is not in the *power* of its algorithms, but in its collection of data; the conditions of the calculation problem have changed. So, consequently, the mechanisms for coordinating behavior will change.

There are two further important changes in the "metamarket," or background conditions, of the neoliberal/cybernetic project. The first is, in some ways, a consequence of its success. The neoliberal extension of markets to ever new domains of behavior has required the *creation* of markets. And, as Nik-Khah and Mirowski have demonstrated, in the mainstream traditions of experimentalist economics and market design, the "ghost of Hayek" lives on in the idea that these are, precisely, "knowledge aggregators" or "information processors." Indeed, not only are markets information processors, but all manner of devices can "be" markets, from platforms to credit cards to surgical procedures.[65]

But the extension of markets and their design raises a second issue: the proper epistemic and computational features of the market were contingent not only on the ignorance of subjects but also their knowledge and rationality. Markets won't do a great job coordinating behavior if people aren't rational. Unfortunately, as the explosion of knowledge in the behavioral sciences has shown since Hayek's time, this is often the case. And, perhaps unsurprisingly, the authors of *Nudge* include one of the world's most prominent behavioral economists. The other has attempted to

---

[64] Sunnstein and Thaler, *Nudge*, 4–8.
[65] See Nik-Khah and Mirowski, "The Ghosts of Hayek in Orthodox Economics," especially 58–62.

articulate a "Hayekian behavioral economics."⁶⁶ The compatibility of Austrian economics, and neoliberalism, and behavioral economics is contentious, but we might note here that Herbert Simon, a major figure in both the history of artificial intelligence *and* behavioral economics, was a student of Lange's but drew on Hayek's account in order to arrive at the concept of "bounded rationality."⁶⁷ As Simon puts it:

> Economics illustrates well how outer and inner environment interact and, in particular, how an intelligent system's adjustment to its outer environment (its *substantive rationality*) is limited by its ability, through knowledge and computation, to discover appropriate adaptive behavior (its *procedural rationality*).⁶⁸

Insofar as the information environment and computational capacities at our disposal have shifted, so has the appropriate adaptive behavior. The point is that by *learning* how individuals *actually*—not just ignorantly, but irrationally or arationally—make decisions, we can *design* markets, or choice scenarios more broadly, *to produce the outcomes we desire.*

It would be nice, I think, if those outcomes were to the benefit of the chooser, perhaps courtesy of a socialist State, rather than to the benefit of the entrepreneurs and shareholders who have decided not simply to make use of the market-AI but to create their own. But in either case, whatever the first-order ends of this behavioral modification, they are governed by a second-order end, namely, to *make sure that our predictions are correct*, by fine-tuning them by continually modifying behavior even further. The feedback loops of platform capital are simply this logic laid bare. To return yet again to the cybernetic metaphors of system theory, what is this but an open system feeding on the information of its complex environment, and returning to it something broken down and simpler?

As Zuboff shows us, we are alienated from our own action under the conditions of platform capitalism, which is in turn rendered as yet more mere behavior. We engage, in order to be incited to engage again: perfect little feedback-driven cybernetic machines.⁶⁹ In the next section, we will

---

⁶⁶ See Sunstein, "Hayekian Behavioral Economics."
⁶⁷ *The Sciences of the Artificial*, 34. See also March and Simon, *Organizations*, 203.
⁶⁸ *The Sciences of the Artificial*, 25.
⁶⁹ For a complementary account, with more detailed discussion of the relationship between governmentality and platform capitalism, see also Gamez, "Inhuman Hermeneneutics of the Self."

explore further how the capacity of these algorithms to produce the truth, and subsequently our rationality, by making us predictable, while at the same time remaining unpredictable to us, fascinates, pointing to the AI invasion from the future. On this, Land was right: contemporary discourse around AI, the irruption of large language models (LLMs) into every aspect of our daily life, confirms the dual dilemma of our technologies making us ever more predictable while themselves eluding prediction.

But before we move on, we should remind Prometheans of another mythical cybernetic figure, namely the great steersman, Charon, who would shepherd souls into the underworld, where they would persist, as a shadow of the forms of life they once led. We should not be surprised that these souls were often brought to Charon by Hermes, the god of communication. It seems that our dynamic systems of information and communication technology do not work toward liberation or emancipation, whether of our desire or our thought, but rather toward the extirpation of human agency entirely.

5.4 No Thyself: Black Boxes at the End of Theory

I have tried to show that the history of AI, from neoliberalism to platform capitalism, is of a piece with the history of cybernetics. Contrary to the Prometheans, rather than simply being a loose coupling, these cybernetic, neoliberal systems share an inner logic or dynamic that serves the sort of "governmentality" described by Foucault. On the other hand, contra the cyberpunk fantasies of the accelerationist, the inhuman AI to which we are submitting is not sexy and chrome but routine and mundane: a beige AI, if you will, that simply maintains behavioral patterns, attempting not to reduce entropy but to pause its increase, momentarily, by perfecting our existence as homeostatic machines. Land is correct that the future is canceled but this is not because AI is "explosive and opportunistic."[70] It is dull and patient, and the only horizon is boredom.

Of course, it is impossible to show that this coupling between capitalist development and the growth of AI is a *necessary* one; maybe, still, things could be different. Perhaps there are technological systems that embody different rationalities, that simply need to be developed and put into conflict with existing ones. Even staunch critics of contemporary AI like

---

[70] Cf. "Teleoplexy: Notes on Acceleration," 511.

Zuboff and Daron Acemoglu take it that there remain different directions for the development of AI.[71] Perhaps the Promethean can cling to that idea. However, in this section I want to claim that, while the Promethean might be able to cling to hope in some destined, or imagined, technological system, some "properly philosophical" artificial intelligence, as Negarestani puts it, rather than the "classically symbolic" kind, to do so is precisely what he refers to as conservative humanism: a conservatism that refuses the radical revision of the human, refusing to *delete* the manifest image, and the ideology of "thought" that belongs to it. And this will become clear once we place the neoliberal/cybernetic project of connectionist AI in the even broader context of technologies of truth-telling.

The Delphic imperative to "know oneself" stands at the inception of Western of philosophy, and if Socrates and Plato inaugurated the investigation of the self, with its various parts and functions, the imperative still echoes through the human sciences today. Arguably, the concepts of our basic commonsense folk-psychology are part of the fruit of that investigation, as are their elaboration in psychoanalysis and computational models of the mind, each of which call into question the authority of some unified coherent self.[72] Sherry Turkle noted the inhumanist elements of both psychoanalysis and computerized thought decades ago:

> People are afraid to think of themselves as machines, that they are controlled, predictable, determined, just as they are afraid to think of themselves as 'driven' by sexual or aggressive impulses. But in the end, even if fearful, people want to explore their sexual and aggressive dimensions; hence, the evocative power and popular appeal of psychoanalytic ideas. Similarly, although fearful, people want to find a way to think about what they experience as the machine aspect of their natures; this is at the heart of the computer's holding power. Thinking about the self as a machine includes the feeling of being 'run' from the outside, out of control because in the control of something beyond the self. Exploring the parts of ourselves that we do not feel in control of is a way to begin to own them, a way to feel more whole.[73]

---

[71] See, e.g., Zuboff, *The Age of Surveillance Capitalism*, 522–523; Acemoglu and Johnson, *Power and Progress*, 323–327.

[72] Foucault of course explored a wide variety of confessional technologies, but consider a different example, in a different register, the production of competing visions of the nature of emotion correlated with different therapeutic technologies for treating them discussed by Richard Sorabji, *Emotion and Peace of Mind from Stoic Agitation to Christian Temptation*.

[73] *The Second Self*, 272.

Both Prometheans and accelerationists are precisely invested in *unleashing* those dimensions of mental life "beyond the self," that escape our control. But the main claim of this section is that the development of contemporary AI, construed as a response to the Delphic imperative, destroys the conception of mind that underlies both.

Neoliberal economics is, alongside psychoanalysis and artificial intelligence, part of this Delphic project.[74] It is not for nothing that, during his own "archaeology of psychoanalysis," which in turn became a "genealogy of *desiring* man," Foucault's attention is drawn to the history of liberalism. Foucault thought that, with the rise of liberalism, the *market* turned into a site of "veridiction," that is, a site in which the "truth" about oneself was revealed. For him, though, knowing the truth about oneself was not necessarily an emancipatory project. He "asked: How had the subject been compelled to decipher himself in regard to what was forbidden?"

... Max Weber posed the question: If one wants to behave rationally and regulate one's action according to true principles, what part of one's self should one renounce? What is the ascetic price of reason? To what kind of asceticism should one submit? I posed the opposite question: How have certain kinds of interdictions required the price of certain kinds of knowledge about oneself? What must one know about oneself in order to be willing to renounce anything?[75]

As we've seen, neoliberal economics *does* construe the market in epistemic terms. But what sort of truths does it reveal? Certainly they aggregate knowledge and communicate information through prices, but they have also remedied our ignorance *about ourselves*. So, for example, economists appeal to "revealed preferences," when market interactions by consumers disclose—at the very least to third parties—their values, desires, and interests. But, of course, as the behavioral economists pointed out, the sorts of inferences required to disclose those preferences demand of economic subjects rationality and no sense of self-control.[76] What is revealed

---

[74] For an extended discussion of this point, see Gamez, "The Place of the Iranian Revolution in the History of Truth." I am not the first to draw this connection. Even those deeply sympathetic to neoliberalism have evoked its resonances with psychoanalysis. See, e.g., Gaus, *On Philosophy, Politics, and Economics*, 1–3.
[75] "Technologies of the Self," 17.
[76] See Thaler, *Misbehaving*, 85–86.

is *one's desire*, which should govern one no matter what.[77] Like Lacan, the liberal mantra is "never give up on one's desire."

Classical neoliberals reject the possibility that anything—whether the state or some sort of AI or platform—could replace the function performed by the market. But they do not all agree on the reasons why. For those in the Austrian tradition, like von Mises, Hayek, and Rothbard, this is because the market was the *only* way, not just to aggregate information, but to disclose the desires or interests of consumers: to reveal their *selves*. Indeed, Hayek's rejection of the possibility of the centralized (State) planning doubles as a rejection of the possibility of surveillance capitalism, "knowledge of the kind which by its nature cannot enter into statistics and therefore cannot be conveyed to any central authority in statistical form."[78] More strongly, for von Mises, we simply do not have *access* to certain forms of self-knowledge without markets, namely, knowledge of our *cardinal* desires; for von Mises, our desires are as close to contentless as possible, mere *ordinal* desires that do not allow for comparison. It is our interactions in the market that transform mere degrees of intensity into comparable content:

> But Mises realized that this insight meant it was absurd to say (as Schumpeter would) that the market 'imputes' the values of consumer goods back to the factors of production. Values are not directly 'imputed'; the imputation process works only indirectly, by means of money prices on the market.[79]

Calculation *produces* cardinal desires. In fact, von Mises and Rothbard think that Hayek misconstrues the entire debate by making it a matter of *knowledge*; the debate is about *calculation*, and we are so unknown to ourselves that a market is necessary simply in order to perform those calculations.

We should note two important upshots of this view. The first, shared by Foucault, the neoliberals, the cyberneticists, the psychoanalysts, the

---

[77] In breaking from techniques of *repressing* one's desire, Foucault suggests that, with the rise of liberalism, from "the eighteenth century to the present, the techniques of verbalization have been reinserted in a different context by the so-called human sciences in order to use them without renunciation of the self but to constitute, positively, a new self. To use these techniques without renouncing oneself constitutes a decisive break" ("Technologies of the Self," 49).

[78] "The Use of Knowledge in Society," 98.

[79] See Rothbard, "The End of Socialism and the Calculation Debate," 65.

Prometheans, and the accelerationists, is that we are *deeply* opaque to ourselves, and we have no authoritative access to our own beliefs and desires. Obeying the Delphic imperative, then, requires us *to be known* by others, capable of interpreting and explaining our behavior. Second, it raises serious questions about the *meaning* of the success of platform capitalism. What would this mean for our self-knowledge, and the self to be known?

There are a few options. First, it could be that the algorithms of platform capitalism come to know our desires, assuming the confessional role neoliberals reserved for the market. As we shall see, however, this seems unlikely; if belief-desire psychology is a theoretical means to predict and order our behavior, the mechanisms of surveillance capitalism dispense with it. So, a second option is that platform capitalism adopts a different strategy for cultivating our self-organization, one that bypasses the need for appeal to beliefs and desires. But were this to be so, it would lead, I think necessarily, to a dramatic revision of our commonsense psychology anyway; if we think that beliefs and desires, or whatever serves as their representational and motivational proxies in however sophisticated a cognitive science one likes, play an important causal role in understanding our behaviors, then the ability of prediction engines like contemporary AI to successfully predict and shape behavior without appeal to them gives us some ground for skepticism. And this is the third option: to *eliminate* these posits from our theoretical repertoire. I will be arguing that this is the only reasonable option for the Promethean, and that their ambivalence between functionalism and eliminativism needs to come down on the side of the latter. Austrians and neoliberals had to strip desire down to almost nothing in order to salvage the necessity of the market against planning, while priests and psychoanalysts needed to track it beyond consciousness to explain our deviations from rationality. In a different context, Quine once wrote that the Humean predicament is the human predicament; in this context, he might be right.[80] Perhaps a genuine posthumanism will be done with both.

To begin to articulate this admittedly pragmatic posthuman eliminativism, we should consider the form of artificial intelligence embodied in platform capitalism, precisely because it is this AI *that knows us*. AI has become the subject of knowledge, and we its objects. We will then see how this impacts the form of "veridiction" or truth-telling, the Delphic imperative under contemporary conditions; in brief, it has become an

---

[80] "Epistemology Naturalized," 72.

interrogation in which there is no self to be known. Against Freud's *wo es war, soll Ich werden*, we find that *wo es war, soll nichts werden*.

Contemporary AI is largely a matter of the deployment of machine learning algorithms upon vast amounts of data. Deep learning algorithms can work, unsupervised, on unclassified data, cutting humans out of the loop, much of which is increasingly high-dimensional, that is, contains an increasingly large number of features to analyze. The combination of high-dimensional data and deep learning algorithms make them, in principle, *alien* forms of intelligence, that is, deeply opaque to us, insofar as they can work to correlate features of data that are not salient, or even accessible, to human beings, using statistical means that outstrip our comprehension. In other words, the path from input to output is an inaccessible "black box."[81] The classic example of this is the now nearly mythical tale of AlphaGo, the learning algorithm designed to challenge human Go masters who had theretofore been unbeaten at the game.[82] Trained on a huge dataset of previous matches, AlphaGo was eventually victorious, though it used strategies that were strange to the point of incomprehensibility to human observers; no "natural" intelligence would have made those moves.

So, we clearly have an example of the sort of alien intelligence that so intrigues the Prometheans and the accelerationists, to which they are so keen to submit. The application of such algorithms to scientific research has enabled a transformation in the way knowledge is produced.[83] Not only does it transform the practice of knowledge production, but it may reveal a reality foreign to human intellect; recently an unsupervised learning algorithm, trained on experimental data from observation of a pendulum, identified an alternative set of fundamental physical variables to

---

[81] For the by-now classic overview of algorithmic opacity, see Jenna Burrell, "How the machine 'thinks': understanding opacity in machine learning algorithms."

[82] For a slightly more detailed discussion of algorithmic opacity that situates AlphaGo in this context, see David Beer, *The Tensions of Algorithmic Thinking: Automation, Intelligence, and the Politics of Knowing*, 115–121. For a tangential but intriguing discussion of AlphaGo in the context of other ethics-political worries about AI, see Vincent C. Müller and Michael Cannon, "Existential Risk from AI and Orthogonality: Can We Have It Both Ways?" Finally, it is worth noting that Peter Thiel, the neoreactionary tech entrepreneur, was a key investor in the founders of DeepMind, the company behind AlphaGo; see McQuillan, *Resisting AI: An Anti-fascist Approach to Artificial Intelligence*, 94.

[83] For a cautiously optimistic perspective on this, see, e.g., Ewen Callaway, "'It Will Change Everything': AI Makes Gigantic Leap in Solving Protein Structures." For a decidedly less optimistic take, see Mirowski, "The future(s) of open science."

predict its motion. Just what these variables correspond to, and what units they would bear, remains opaque.[84] All of this is to say that machine learning not only evinces intelligence without intelligibility, but points toward alien ways of understanding the world it inhabits. Prometheans, with their distinction between the manifest and scientific images, should welcome minds that grasp the world with new concepts (or perhaps without concepts at all).

More recently, these algorithms have been put to the test of producing intelligible human language; several large language models (LLMs), trained on a massive corpus with billions upon billions of parameters, have been developed over the last few years, obliterating the Turing Test, and serving as fascinating conversation partners and untrustworthy search engines. But the mechanism is the same. From the inputs of linguistic behavior, machine learning algorithms make predictions about what word or utterance is most likely to come next, and in doing so produce largely intelligible discourse: at least, as largely intelligible as most human discourse.

At any rate, at this point there seem very few markers of intelligent behavior that cannot be captured by the algorithm. And no posthumanist ought to rule out the possibility of algorithmically producing some sort of behavior *a priori*. As with any technology, we are likely to continue outsourcing our action to machines (that we once thought were merely our tools) until, beyond the data imperative, we face what Orit Halpern and Robert Mitchell call "the smartness mandate":

> … the demand, cast by its advocates as having the force and irresistibility of a law of nature, that all social processes become smart. A social process becomes smart when the *populations within which that process occurs are redesigned as experimental zones, so that widely distributed forms of electronic sensing produce data that can be processed algorithmically, and in this way enable constant…adaptation…*. [T]he smartness mandate seems to emanate less from sites and structures of human government than from nature itself and its evolutionary processes.[85]

Though, like Zuboff and the Prometheans, Halpern and Mitchell want to argue *against* the inevitability of this mandate, the coupling of data

---

[84] Chen et al., "Automated discovery of fundamental variables hidden in experimental data."
[85] *The Smartness Mandate*, 219.

collection and the predictive, responsive government of action by algorithm seems inescapable at this point. And once again we encounter little homeostasis engines. LLMs, and other similar algorithms, as they colonize various areas of human action, work to produce *the normal, the expected, the predictable*: as we engage with them, we should expect our language use to become less innovative, our queries to become standardized. The behavior won't stop, but it will be looped.

What is important, here, is that we have intelligence without reason, or without thought. Land would not be surprised by this, no doubt, but it certainly speaks against the Promethean project, which Negarestani describes in forceful terms as articulating the attempt

> to let intelligence cultivate itself by turning ourselves into the history of intelligence rather than founding ourselves as its nature or as a totality that must be preserved as the object of its commemoration or striving.... Any system of thought that has a problem with what intelligence does to itself in order to remain intelligent and intelligible is an unfortunate historical phenomenon, and certainly will not be around for much longer.[86]

But on this view, intelligence is seen as implying *reason*; no genuine intelligence without sapience. That is, intelligence is not fundamental but the *result* of mental activity. But can we still unproblematically hold this view? Don't the Prometheans, in fact, have a "problem with what intelligence does to itself in order to remain intelligent"? Don't they remain trapped in an outdated theory of intelligence that chains it to the invisible fetters of thought?

Some might point to the fact that LLMs are "stochastic parrots" or "bullshit engines" and that this looping of behavior will degrade some sort of intrinsic human creativity.[87] But we have already pointed out that the increasing predictability of human beings is part of the entire project; it remains merely an article of humanist faith that there is something "intrinsically creative" about human being. As for parroting bullshit, these concerns are, by and large, yet another version of Leibniz's Mill, what Raphaël Millière and Cameron Buckner call the "Re-description Fallacy."[88] The point is that the description of the material mechanisms of cognition

---

[86] *Intelligence and Spirit*, 492–493.
[87] See Hicks, Michael Townsen, James Humphries, and Joe Slater, "ChatGPT is bullshit"; Damien Patrick Williams, "Bias Optimizers," and Shannon Vallor, *The AI Mirror*.
[88] "A Philosophical Introduction to Large Language Models," 9.

might take place on a different "level" than the description of those cognitions themselves. Prometheans should here recall Sellars; a description of the "real order" might bear no resemblance to the description of the "logical order" of the mind.

It is here that the Promethean ambivalence between functionalism and eliminativism can be resolved. Sellarsian functionalism is *internal* to the manifest image. In the same way that the contrast between, say, "spewing bullshit" and "making conversation" is internal to a description of behavior that appeals to norms of truthfulness, rationality, etc., so a functionalist construal of reason, or agency, or cognition more generally is an internal revision of the manifest image, rather than a reconciliation between the manifest and scientific images. Indeed, even in those cases where Prometheans engage in depth with computation, they view it, largely, as a matter of the symbolic manipulation by means of structure-sensitive rules, that is, in the terms of symbolic or "good old fashioned AI" or "GOFAI."[89] It is precisely this viewpoint, for instance, that leads Negarestani to articulate a "computational functionalism":

> ... [t]he mind is what it does. While this mental or noetic doing can be taken as constrained by the structural complexity of its material substrate, it should be described in the functional vocabulary of activities or doings. The mind—be it taken as an integration of distinct yet interconnected activities related to perception, thinking, and intention or seen as a cognitive-practical project whose meanings and ramifications are still largely unknown (à la Hegel and Mou Zongsan)—has primarily a functional import.[90]

Like Christias and Wolfendale, Negarestani thinks that the mind "is what it does," and thus that radical functionalism can depart from a humanist vision of intelligence by, following Turing, recognizing that the mind can be "multiply realized" on different material substrates. Provided that these material substrates can be organized to perform computational functions, the capacities of (cognitive) agents can be vastly expanded. But their view relies on a GOFAI-inflected, symbolic, or manifest, view of computation. It ignores the fact that contemporary AI is, by and large, connectionist, and that there may well be *nothing* in the scientific image

---

[89] See Margaret Boden, *Mind as Machine*, Ch. 10. As Boden points out, the cybernetic, connectionist vision was usually "fiercely opposed to GOFAI" (13).

[90] "Revolution Backwards: Functional Realization and Computational Implementation," 140–141.

that can be accurately described as performing these sorts of functions. Whatever relationships obtain between the activation vectors of neurons, they are not the normative relationships that obtain between assertible contents.

While early eliminativists like Richard Rorty did not engage at any length with AI, what might be called the second wave of eliminativism, as developed by Paul Churchland, Stephen Stich, and others, drew heavily on the development of connectionism.[91] It is striking to note that the Prometheans have not engaged at any length with the development of connectionist AI, despite the fact that, as a model of mind, it has overcome many of the limitations that plagued it during previous AI summers.[92] They seem to remain "chained" to a conservative image of thought.

As we saw in Sect. 4.1, Sellars' distinction between the "manifest image" and the "scientific image" is a distinction, at least initially, between *two theories*, or broad theoretical frameworks. With respect to the theoretical elaboration of human behavior, the "manifest image" developed without access to the internal structure of the human being,—modeling instead those internal states—thoughts, beliefs, desires, etc.—on what publicly accessible and available data there was: verbal speech and action. The "scientific image," on the other hand, begins from that internal neurophysiological structure, and works outward. Sellars, of course, believed the latter was destined for victory. But what matters, for our purposes, is that in principle, we are in each case confronted with a phenomenon that *calls* for a sort of explanation; for lack of a better term, we confront the mind as a "black box," and thus when we aim at explanation we aim at "filling in" that missing internal structure.

By this I don't mean that, in Heideggerean language, we encounter the world phenomenologically in its *vorhandenheit*. It seems likely that we are initiated into practices of encountering what is that are always already couched in the language of the manifest image, such that we primarily confront the black box only when that image collapses, when our figurative or literal hammers break. But the Sellarsian point is that, when that happens, the scientific image is authoritative, and the manifest image, *qua*

[91] See Rorty, "In Defense of Eliminativism," as well as Brandom, "An Arc of Thought: From Rorty's Eliminative Materialism to His Pragmatism." For the second wave, see Churchland, "On the Nature of Theories," Ramsey, Stich, and Garon, "Connectionism, Eliminativism, and the Future of Folk Psychology," Stich, "From Connectionism to Eliminativism."

[92] See Millière, "Philosophy of cognitive science in the age of deep learning."

image, is revealed as a long-standing, but deficient alternative. There are two points to be made here. First, from the Sellarsian/Promethean view, it is not at all clear why, if we think that one can salvage the language of the manifest image or stance, the category of "person" should not include AI; as good enlightened posthumanists, recognizing that agency and intelligence are as agency and intelligence do, we must begin from its intelligent behavior.

The second point was actually made decades ago, by Daniel Dennett, namely, that treating a system, whether biological or technological or, frankly, a cyborg, whose behavior is in need of explanation, *as an intentional system* is a stance that one takes toward it as part of a more or less pragmatic "predictive strategy."[93] That is, to place a subject in the manifest image, as an inhabitant of the space of reasons, is part of an attempt to *predict their behavior* via explanation. The prediction of behavior, of course, is the central goal of platform capitalism and, more broadly, of attempts to manage, manipulate, and control behavior.

On this view, *thought*—the realm of belief and desire—is a theoretical posit or postulate that can serve in the explanation of intelligent behavior, through the application and enforcement of a constitutive ideal of rationality. But intelligence comes first, and then we fill in the black box. Alongside intentionality, Dennett describes the "physical stance" and the "design stance," predictive strategies for systems to which we have access to the internal dynamic structure whose physical relations we can track, and for systems to which we can attribute a purposive mapping from inputs to outputs. He even describes computers, precisely, as particularly apt for the design stance.[94] And, of course, we absolutely *can* adopt the design stance with the machine learning algorithms of contemporary AI; it precisely allows us to "black box" their internal processes. But, insofar as we seem to have already welcomed AI into the circle of intelligent beings, we want more than this. We want to hold them accountable. And this accountability is, similarly, a mechanism by which we can regulate behavior, in order to make it better fit our predictions. And adopting the intentional stance is a wonderfully effective way for us to do so.

Placing a subject in the space of reasons—adopting the intentional (or manifest) stance toward them—doesn't just *explain* in order to make *predictions*, it also *sanctions and regularizes*. We hold those subjects accountable to standards of rationality, and in doing so shape them into subjects

---

[93] Dennett, "True Believers: The Intentional Strategy and Why It Works," 59.
[94] "True Believers: The Intentional Strategy and Why It Works," 60–61.

of reason. We know that no human comes into this world as a rational creature, and we know further that initiating them into the practice of reason is a social practice of shaping behavior to accord with the norms of thought. Nietzsche called this the "making predictable" of humanity.[95] In this sense, the history of rationality is recapitulated in the prediction-intervention feedback loops of platform capitalism.

Because AI clearly exhibits intelligent behavior, to the point where we outsource not just mundane acts but cognitive tasks, while at the same time producing outputs through procedures that can seem deeply foreign to us, we similarly want to *hold the machines accountable*, to make them trustworthy. This impulse arises, naturally enough, from concerns about algorithmic bias, injustice, and inequities that result from, e.g., the use of problematic training data or the workings of opaque algorithms themselves. If the physical stance were sufficient to *place checks* on their outputs, we wouldn't have to worry. But deep and persistent algorithmic opacity, arguably an intrinsic feature of the complexity of the neural nets that power AI, prevents us from adopting it straightforwardly. These concerns have been behind the explosion of interest in interpretable or explainable artificial intelligence (XAI). XAI is a project of making accountable.

Luckily for us, the intentional stance is a particularly successful strategy for some systems, which is appropriate to take up *precisely when we don't have access to others*, when we cannot attribute an overall purpose or function or grasp the underlying physical (or, in this case, informational) dynamics. And so we should not be surprised to find that proposals to adopt the intentional stance occupy an important place within XAI discourse and related psychological literature around human–machine interaction.[96] It seems appropriate even in those cases *in which we know* that the internal process does not match or mirror the external output; the intelligibility of the explanation, and its rough fit, is worth the cost of accuracy

---

[95] *On the Genealogy of Morality*, 36.

[96] Zerilli, "Explaining Machine Learning Decisions," Zerilli et al., "Transparency in Algorithmic and Human Decision-Making: Is There a Double Standard?", Günther and Kasirzadeh, "Algorithmic and Human Decision Making: For a Double Standard of Transparency," Kasenberg, Arnold, and Scheutz, "Norms, Rewards, and the Intentional Stance: Comparing Machine Learning Approaches to Ethical Training," Thellman, de Graaf, and Ziemke, "Mental State Attribution to Robots: A Systematic Review of Conceptions, Methods, and Findings," Perez-Osorio and Wykowska, "Attributing the Intentional Stance to Natural and Artificial Agents," and Marchesi et al., "Do We Attribute the Intentional Stance to Humanoid Robots?"

and realism. Even those who recommend a healthy pluralism of approaches to XAI acknowledge a place for the intentional stance, on the grounds that it works well enough for similar complex physical-information systems, namely, humans.[97]

The point to make is that *thought*—the attribution of beliefs and desires knitted together through rationality—is a *tool*, an instrument developed to make sense of the behavior of machines we already recognize as sufficiently like us as to demand accountability. But reason and thought do not supervene on any internal state of the machines we hope to explain and predict any more than they supervene on ours; they are a *mask* that we can—perhaps must— place on the machine to explain them, to domesticate their alien intelligence, in the recognition that what's actually going on outstrips us.

In its very applicability to machine learning algorithms, the intentional stance reveals itself as a pragmatic attitude, which makes possible an ethical orientation; we are able to *face* each other, and the machine, because we adopt it, that is, because we give each other faces, make ourselves sources of *expression*, manifesting an invented interiority.

Unfortunately, this generosity in extending the facade of mind to the machine is not reciprocal. Antoinette Rouvroy makes the point quite well, discussing what she calls "algorithmic governmentality":

> Of course, algorithmic governmentality permits to remove all... encounters: appearance before the law, avowals, testimonies.... Foucault describes the avowal as being exactly what the individual uses to become what he or she is, to become what he or she has done. This extremely subjectivising aspect of avowal, of course, we no longer find it... There is no moment for subjectivation, there is no interpellation other than interpellations by machines, by profiles and these are not really interpellations since on contrary, the machine answers in your place.[98]

The surrender of thought is asymmetrical. In this case, it is the accelerationists who are half right. While we adopt the intentional stance toward, that is, bestow the human mask of rationality upon, the system of algorithms that, collectively, constitute the infrastructure of platform

---

[97] Zednik, "Solving the Black Box Problem: A Normative Framework for Explainable Artificial Intelligence," 269.
[98] Rouvroy and Stiegler, "The Digital Regime of Truth: From the Algorithmic Governmentality to a New Rule of Law," 15.

capitalism, these systems ruthlessly work to strip from us the need for masks entirely. AI invades from the future, removing our human face, except we are not confronted with a robotic face beneath, but no visage at all.

In *Wired* magazine in 2008, Chris Anderson announced "the end of theory" in the wake of the Big Data revolution, the same revolution that provides the material on which contemporary AI runs and works.[99] In the briefest terms, the idea—as with Dennett—is that the *point* of our scientific theories, our frameworks of laws, ontologies of objects, models relating them, is to explain and consequently *predict* observable phenomena.[100] These scientific theories are necessary because our data is limited, and thus too is our ability to correlate phenomena and make inductive predictions. Theories support counterfactuals, providing us with mechanisms and postulates to delegitimate spurious correlations, and so on.[101] But, Anderson suggests, these are now obsolete; we have—or will have—sufficient data, and have—or will have —sufficiently sophisticated algorithms, allowing us to simply make predictions based on what Sellars called "correlational induction."[102] Indeed, in this respect AI obeys the constraints of the manifest image far better than humans have. And if science aims at predictive accuracy, then we have finally reached a point where correlation will be sufficient.[103]

---

[99] "The End of Theory: The Data Deluge Makes the Scientific Method Obsolete," *Wired*, June 23, 2008. https://www.wired.com/2008/06/pb-theory/.

[100] In this, it shares in the basic attitude of van Fraassen's constructive empiricism. See *The Scientific Image*. Of course, van Fraassen takes his title from Sellars, and part of the aim of his work is precisely to argue that the scientific image is just that—an image, not reality—and the manifest image isn't an image at all.

[101] Gary Smith and Jay Cordes object to Anderson's claim on the basis that Big Data will often detect "phantom patterns" with either no predictive value or no objective validity. But we might think that this is just a contingent problem of the data and algorithms used, rather than a principled objection; it is perhaps telling that they begin their chapter on "the paradox of Big Data" with an example of the failure of the computational analysis of language, something with which recent LLMs have had absolutely massive success. See *The Phantom Pattern Problem: The Mirage of Big Data*, 101–102.

[102] "Philosophy and the Scientific Image of Man," 7. Wolfgang Putsch calls this a "new inductivism" and attempts to reframe concepts like "causation" in its terms. See *On the Epistemology of Data Science: Conceptual Tools for a New Inductivism*.

[103] This has, of course, been controversial; see, e.g., Pigliucci, "The end of theory in science?" However, even the more forceful arguments against Anderson's thesis are strikingly sober. So, e.g., Mazzocchi essentially just urges caution in abandoning the traditional scientific aims of (model and theory-driven causal) explanation ("Could Big Data be the end of

The unsettling asymmetry becomes clearer when we realize that the end of theory also means, in some sense, the end of the manifest image *for us*; the theoretical posits of thought and rationality, of beliefs and desires, representations and motivations that were needed to explain *our* intelligent behavior can be replaced by mere correlation. In other words, AI, as the system that produces the truth about *who we are*, the truth-teller at the end of the Delphic project, which has always been just as much about subjecting ourselves as about discovering ourselves, will be able to *tell the truth about us without theory, that is, without internality*.

The liberal unleashing of desire gives way to a normless incitement of behavior. The surplus data we provide in our engagements across all platforms—which is just to say, here and now, in all of our behavior—can be analyzed to make predictions, which in turn make possible ever more arational interventions upon us, increasing the accuracy of prediction, and so on. There is no need for theory at all. Even if we reject it ourselves, the machines have reached the end of theory: for them we are simply black boxes, into which the beliefs, desires, rationality, those theoretical posits that we attributed to ourselves in the earliest phases of the process of making ourselves predictable, dissolve. Contra *both* the accelerationists and the Prometheans, against the partisans of desire and the comrades of reason, AI overcomes the contingencies making interpreting our behavior in terms of either belief or desire seem necessary. Our human faces are, indeed, removed.

What to make of this, from the Promethean point of view? What has become of intelligence and, in turn, what has intelligence made of us? Maybe, as Negarestani hopes, there remains some substantive, normative, revisionary rationality embodied in the commitment to humanity as a technological project. But if so, it's not clear what access we have to it. We can certainly view the manifest image as technology, as a tool for the construction of the space of reasons. But what we see is that this aim is part of the twin projects of rendering accountable and making predictable, and thought, belief, desire, reason, etc. were mere instruments that give way to others in serving those goals. When we were alien, inexplicable, unpredictable, to ourselves, we adopted them; the alienness of AI demands

---

theory in Science?"), while Cabrera adopts the defensive position of arguing that causation and explanation may not be dispensable to scientific practice ("The Fate of Explanation in the Age of Big Data"). However, all of this leaves open the possibility that what science is, and what we can hope to gain from it, is in the process of transformation.

them from us now. But it does not demand thought from us as AI disabuses us of the need to make use of it; if our cognition can be outsourced, our every activity can be recorded, analyzed, and correlated to produce accurate predictions of our behavior, feeding back into an ever-tightening loop; we have no use of the manifest image. There is simply *nothing* going on.

Before concluding, I want to briefly address a concern that one might have about the image I have presented of the end of the Delphic project, and the exhaustion of subjectivity that it portends. For do we not still have an interest in *explaining* ourselves and the world we live in, and in being understood? The adoption of the intentional stance, the development of the manifest image, might have purported to give us a causal grip on our behavior, allowing us to govern ourselves and others, but it also was a tool for some sort of self-understanding. Understanding our minds might be valuable in itself, possibly even a source of resistance against various forms of oppression.[104] But the exhaustion of subjectivity, the effacement of agency, insofar as it erases the distinction between behavior and action, gives up on anything we would consider explanation. This might give us some reason to *resist* it by holding on to the explanatory framework of the manifest image.

But if we are in need of explanations, it is not clear why we would need to revert to the manifest image. For Dennett, for example, the intentional stance is appropriate when the design stance and physical stance are unavailable, or are not sufficiently developed, such that intentional stance offers better predictions. If mere correlational induction is more effective, why bother? We know it is not *literally* an accurate representation of the world, insofar as we know that neither the manifest nor scientific images are "true," in any traditional sense. But the scientific image, at least, would have the virtue of "picturing" the world properly. However, it is not at all clear that the concepts of the scientific image would be relevant; the explananda—our actions—have no place here, are not part of the furniture of the world.

Indeed, here we might question general demands for explanation as such, especially in terms of theoretical posits, or causes, which "beliefs"

---

[104] I leave aside here any technical distinctions between "explanation" and "understanding," such as one might find in methodological disputes in the human sciences. In the very broad sense in which I am using the term, to explain something is to have some sort of understanding of it.

and "desires" would have to be. Here I appeal to another Sellarsian. Bas van Fraassen has argued precisely that "There cannot be a requirement upon science to provide a theoretical elimination of coincidences, or accidental correlations in general, for that does not even make sense."[105] The sort of explanatory demand he criticizes is a demand that correlated phenomena—what he calls "the appearances," and in our context would include, say, the environmental inputs and behavioral outputs tracked by platform capitalism—be explained in terms of a "common cause," some further, perhaps hidden, factor. He demonstrates that the phenomenon of quantum entanglement should lead physicists, *on physical grounds*, to abandon the idea that correlated phenomena will always have an explanation in terms of real, local hidden variables, that is, unobserved theoretical posits functioning as common causes.[106] There are real, genuine correlations between entangled particles that cannot share a common cause. More generally, he refutes "the presupposition that genuine probabilistic dependencies always arise from causal relations."[107] That is, our best physics indicates that that demand for explanation hits bedrock in brute facts. While the examples come from physics, the point generalizes. There may be, simply, causal gaps, the same non-causal void discussed in Sect. 1.2 with respect to the statistical mechanical interpretation of the Second Law, information, and entropy. I think we should not be surprised to find that the cybernetic imaginary, the cybernetic project, which began precisely in modeling the intentions, that is to say, the internal causes, of enemy pilots on probabilistic feedback loops should end up with simply those loops.[108]

I cannot help but hope that the irony of appealing to van Fraassen here is appreciated. Sellarsian though he may be in many respects, his articulation of the limits of the demand for explanation was intended precisely to preserve the manifest image in the face of the imperialist ambitions of scientific realism, which "is always a search for hidden variables."[109] By salvaging the possibility of regions of the lifeworld to be handled by mere "correlative induction," without the use of theoretical posits, we could

---

[105] *The Scientific Image*, 25.
[106] See "The Charybdis of Realism: Epistemological Implications of Bell's Theorem." I leave aside other technical examples that challenge, at the very least, the broad ideal of "causal" explanation, such as Norton's Dome. See Norton, "Causation as Folk Science."
[107] Hausman, "Lessons from Quantum Mechanics," 90.
[108] Galison, "The Ontology of the Enemy," 238–245.
[109] van Fraassen, "Rational Belief and the Common Cause Principle," 207.

"save the appearances," maintain a sense of enchantment. We should hardly be surprised if the unpredictable algorithms who seem much more worthy of being engaged from the intentional stance come to view *us*, as the subjects of their knowledge, about whom they aim to "tell the truth," in the most paradigmatic terms of the manifest image.

From Delphi to Vienna and Mont Pelerin to MIT and Silicon Valley, the project of knowing ourselves has involved submitting to authorities that would explain to us why we do the things we do, in terms of what is going on inside, but there is nothing left in there. Having been emptied out, the history of our truth-telling about ourselves ends not with a confession but a whimper. Indeed, without the notions of belief and desire, it is not entirely clear that the concepts of agency, subjectivity, or selfhood survive. Our own self-image needs to be revised, in the light of intelligence, but not as either the Promethean or the accelerationist envisions.

CHAPTER 6

Concluding Remarks: Defacing Thought, or, on Being Disoriented

Abstract In this final chapter, I summarize the results of our exploration of Prometheanism, gesturing at the bleak, directionless posthumanism that I think would be the result of taking the history of AI, and its continuity with the project of neoliberal, cybernetic self-governance. I once again (briefly) contrast my work here with that of Bernard Stiegler, before suggesting that recent work on the concept of "the face" might intimate something of the faceless future of the post/human, in a world in which it makes more sense to think of AI as minded than ourselves.

Keywords Face • Future • Posthumanism • Negentropy • Alien • Machine intelligence

I have tried to argue that Prometheanism, as a posthumanist, rationalist strategy or project, has been unable, despite its best efforts, to break with its accelerationist antecedents. And this dooms it to failure. Both hinge on a commitment to some posited, sub- or transpersonal—one might say "sub-" or "pre-human," were it not for the connotation—component of our mental lives, which might be freed from its fetters by the explosion of machine intelligence: reason, for the Promethean, and desire, for the accelerationists. But this attempt to place the inhuman core of the mind

© The Author(s), under exclusive license to Springer Nature
Switzerland AG 2025
P. Gamez, *Posthumanism Meets Surveillance Capitalism*, Palgrave
Studies in the Future of Humanity and its Successors,
https://doi.org/10.1007/978-3-031-90770-8_6

within the dynamics of our technological systems—the invasion of artificial intelligence from the future—collapses. The mind falls into ruin, becoming a mask for AI and one that humans no longer have any need for.

I attempted to frame this, as many accelerationists and Prometheans do, in the terms of science fiction, drawing on the cybernetic imaginary of Asimov's "The Last Question," before situating their project in a very real cybernetic, and neoliberal, historical trajectory. In Asimov's story, the last question is "Can entropy be reduced?" Colloquially, this is a way of asking: Can we reduce disorder? Can we *order* the universe, our polities, our selves, against an implacable, impersonal process that marches ever forward? Ironically, from the human point of view, it is this disorder that is most predictable; not chaos but equilibrium, without structure. These themes of entropy, information, and prediction showed up throughout; the aim of maintaining some degree of predictable order, that is, some form of homeostasis, repeats again and again.

I opened this book by echoing Brassier's question: What is it to *orient* ourselves toward the future? Given the thermodynamic arrow of time, what should we do with the position in which we find ourselves? In Asimov's text, it's already here: artificial intelligence reaches back through time, enacting creation in reverse. Accelerationists and Prometheans think the same. We can, like the former, claim that AI ruins our dreams of rationality, inciting and intensifying our desires beyond the control of reason or, like the latter, claim that it is constructing itself through us as an intelligent technological system. Or, we can recognize that it is doing neither, as we surrender thought to it while defacing our own.

In some ways, the account I have presented is similar to that presented by Bernard Stiegler in his later works. He too is concerned by the way an "industry of traces," feeding off our behavioral exhaust, is kicked into overdrive by the rise of digital technologies that constitute our "hyperindustrial society [that] is fully accomplished as the automatization of existences."[1] He even refers to Anderson's "end of theory" thesis, which he takes to be the epitome of liberal epistemology (though we might think it represents something like the end of epistemology).[2] Stiegler thinks about this process of "automatization" as a matter of metaphorically increasing entropy, which could in principle be combatted by increasing "negentropy":

---

[1] *Automatic Society*, 20.
[2] *Automatic Society*, 45.

6 CONCLUDING REMARKS: DEFACING THOUGHT, OR, ON BEING... 133

The Anthropocene is an 'Entropocene', that is, a period in which entropy is produced on a massive scale, thanks precisely to the fact that what has been liquidated and automated is knowledge, so that in fact it is no longer knowledge at all, but rather a set of closed systems, that is, entropic systems. Knowledge is an open system: it always includes a capacity for disautomatization that produces negentropy. When Chris Anderson announced that the era of 'big data', or what he calls the 'data deluge', would lead to the 'end of theory', he made a serious mistake, given that he ignored the fact that to close an open system leads in a systemic way to its disappearance.[3]

Of course, this view hinges on the idea that "knowledge"—whatever, exactly, he means by that—is an open system that needs to feed on the negentropy of its surroundings. He neglects the possibility that it is, rather, platform capitalism that is the open system, feeding on *our* information, leading not to its own disappearance but ours. Platform capitalism has answered Brassier's question "what should we do with time" by being *done* with time, looping it endlessly. Unlike Asimov's AC, the point is a sort of *stasis*, an attempt not to decrease entropy, to reverse it, to redirect time, but to halt it. And if it is the Second Law that gives rise to the arrow of time, which is to say, temporal orientation, then to freeze it is to *dis*orient ourselves, leaving us without direction, facing nowhere.

Stieger's talk of "disappearance" evokes what Deleuze and Guattari referred to as "becoming-imperceptible," a way of escaping from the regimentation of our political institutions. And, coincidentally, this becoming-imperceptible occurs in a chapter of *Thousand Plateaus* entitled "Year Zero: Faciality." For Deleuze and Guattari, disappearing is a way to "dismantle" the face:

> That is why we have been addressing just two problems exclusively: the relation of the face to the abstract machine that produces it, and the relation of the face to the assemblages of power that require that social production. The face is a politics.[4]

For them, *having a face* is also a matter of power; the fact that we humans have a membrane, an interface that is supposed to *express* our inner states, makes us legible to others and subjects us to codified rules of signification and behavioral norms. Recent scholars have discussed, for

---

[3] *The Neganthropocene*, 51–52.
[4] *A Thousand Plateaus*, 181.

example, precisely how algorithmic facial recognition technologies make us available to surveillance and manipulation. It is as if being *recognized*, while a condition for respect, were at the same time the condition of being targeted, constrained, and channeled.[5] We can see how Land would be interested in accelerating the development of abstract machines to free our desires from this need for recognition.

If the face can be dismantled, then it is contingent. Carl Olsson has recently explored the contingent evolutionary history of the face, from rudimentary clusters of light-receptive nerves in distant ancestors to the sophisticated faces we seem to bear today. Even further than Deleuze and Guattari, Olsson describes not only how the face has led us to be certain kinds of subjects but, in giving us the possibility of *facing* a direction, a person, a situation, has likewise determined not only how we perceive but how we conceive of thought, personhood, and so on. The face is a condition of our being oriented, but one perhaps nearing its end, precisely as we make the world ever more intelligible to alien intelligences whose machinic existence has never been oriented toward its environment in the same way:

> The face began in the unity of directed motion, developed into a sensory composite that enabled subjective awareness in the process of distinguishing one's own body from the environment, and eventually coalesced into a social operator in a small number of lineages which have had transformative effects upon almost all terrestrial habitats. What we are now seeing is that other distributions of perceptual 'organs' and other structures of information processing are taking on roles that faces had previously captured. To understand this is to understand that the face's role as a driver of articulating the noosphere may be nearing its end. The human, at the end of the day, may well become known as the species that defaced the world.... The new arrangement of the noosphere that has been created to suit faces need not resemble how machinic intelligent systems would organise their environments at all. We know that the physical arrangements of the planet's surface are being remade to suit the sensing, modelling, and acting that is most efficient for machine intelligence.[6]

Olsson, like a good Promethean, is sanguine about this state of affairs, seeing in it the possibility of emancipating thought from its parochial fetters:

---

[5] See Celis Bueno, "The Face Revisited" and Olsson, "Peak Face."
[6] "Peak Face," 15.

If facing forward has impacted our conceptual repertoire and our civilisational history on a fundamental level, perhaps this post-peak-face *homo sphera* would realise a faceless democratic ideal equally accommodating to all directions of thought.[7]

But Olsson doesn't seem to consider the possibility that, rather than being accommodating to all directions, this "faceless ideal" is accommodating to none. That is, as the world becomes intelligible to machine intelligence, it may be in the process of removing any coordinates for ourselves. Whatever the destiny of thought is, there is nothing to guarantee that we share in it.

In his early, arguably pre-Promethean work, Brassier considers this very possibility. The climax of *Nihil Unbound* consists in a discussion of the philosophical significance of extinction and, more specifically, the temporality of thought. More than the mere disappearance of species, or the death of the sun, discussed by Lyotard, which would demand thought go on without the basic substrate of organic life, *heat death*, or the total reign of entropy, constitutes the erasure of all thought's horizons. *That* such extinction will take place makes the universe a place in which thought is not a transcendent representation of reality, made possible by temporal projection, but one in which thought is an object like any other. In that work, it was not Cosmic AC, the divine artificial intelligence, that worked backwards in time but rather "Extinction [that] seizes the present... between the double pincers of a future that *has always already been*, and a past that is *perpetually yet to be*."[8] The posthuman predicament of contemporary platform capitalism places us in a related position; thought is an object among others, though not rendered so by the fact of extinction but by the realization that thought was a mask all along, and our business with its future is not a matter of any particular link between it and either humanity *or* its posthuman descendants. Whichever entities aim to reduce or maintain their entropy, whichever sources of negentropy are fed upon to do so, is completely arbitrary.

So perhaps "How we should orient ourselves toward, or face, the future?" is the wrong question. For the Promethean, putting the human into question, opening ourselves to its overcoming, is also to raise the question of radical futurity. This contrasts with what David Roden calls "critical posthumanism," which largely attempts to show that the

---

[7] "Peak Face," 16.
[8] *Nihil Unbound*, 230.

conceptual distinctions between humans and various nonhumans—whether animal, technological, divine—are unstable, themselves an artifice or project of constructing a human that we have never been.[9] For such views, understanding the production of the human requires a genealogical vision which loosens the conceptual grip of the human and opens up a variety of futures, not a rupture opened up by a future arriving to radically transform "us," in some sense of "us" yet to be determined. For the post-Landian Promethean, we confront futurity as a matter of thought and its transformation. But I contend that once AI peels back our human faces, there *is* no facing the future, because there is no *facing* anything, at least, not before we decide what sort of mask we are wearing.

The accelerationist concern for the future is, after all, an attempt to let an Outside in, to find an exit from the limits of the human condition. And the direction from which that Outside enters, or in which we exit, is futural, is futurity itself. But—I have tried to suggest—this is deeply limiting; the Outside is already within us, and the exit is in every direction. Put differently, this metaphor of exteriority and interiority misleads, and the face, as interface, is just as wrapped up in that confusion. The manifest image of the human being, governed by reason and driven by desires, is just a mask that we have worn, and now one that is bestowed to artificial intelligence that can see us clearly without it. Our interiority was a mask the whole time; not in the sense that it *hid* something but rather in that *it gave us a face*, and thus made it possible for us be oriented to begin with.

But there is nothing about the way the world is, nothing about the way things are, that forced us to don the mask of reason. Indeed it is an artifact of error or failure, but productive nonetheless; a way of being wrong about ourselves, and coping with just how alien we are to ourselves. And just as we productively misrecognized ourselves, we can productively misrecognize others. AI is hardly the first nonhuman that has been welcomed into the circle of persons. There are many rich traditions of animism; humanism is but one of them. So, rather than asking what we might do with time, or how we might become alien to ourselves by submitting to inhuman reason or opening ourselves to a radically foreign future, we might rather recognize that the alien/non-alien and human/inhuman/posthuman distinctions were not the most important or interesting to begin with.

---

[9] See *Posthuman Life*, Ch. 1. Roden focuses on post-deconstructive thinkers like Hayles and Braidotti. But beyond this selection, for example, we have already seen how Andy Clark describes humans as "natural born cyborgs" (Sect. 4.2), and we might refer also to Agamben's "anthropological machine" producing human difference (cf. Agamben, *The Open: Man and Animal*).

If "*Humani nihil a me alienum puto*" has long been the expression of a certain ethic of humanism, we posthumanists, sadly, might have to admit that, rather, nothing alien is alien to us. And insofar as this is true, we find ourselves disoriented: we cannot give ourselves over to the alien, and yet we are not at home with ourselves, have no nature to be expressed. In contrast to the question of Promethean futurity, we find ourselves with less of a question than a decision to be made within an endless present. Not "what is to be done?" but "where and who are we?" Not a matter of *discovering* but a matter of inventing, of placing ourselves within any number of histories and open to any number of histories.

The thrust of this book has been largely exegetical and critical, pointing out the consequences, given the framework of Prometheanism, of platform capitalism. In these concluding remarks, I have only been able to gesture at the shape of a posthuman reckoning with those consequences. And it is, in many regards, bleak. There is perhaps some small comfort to be found in the fact that the situation confronting the post-accelerationist is, in a sense, the realization of the project. Not that some subpersonal aspect of mind has been unleashed such that its trajectory might carry us beyond the confines of the human. But rather that an even more radical sort of freedom has been achieved. In wriggling out from under the weight of the human mind, we have disburdened ourselves of the necessity of being oriented by its destiny or its purported relation to the world, or Being, or what have you, or its relentless functioning in the deferral of entropic breakdown. The ways in which we invent who we are and how we ought to live—even the choices we make about what it is to choose, or act—are absolutely groundless, no more indebted to a past and open to a future than vice versa. There is no hope that in becoming either gods or monsters we avoid the disorientation that is our lot, human or not. But if to put on a mask is to present oneself to another, to orient the present toward another, then at least we are not, and cannot be, in all our terrible freedom, alone.

REFERENCES

Acemoglu, Daron. "Hayek vs AI Socialism". *International Economy.* 37:2 (2003): 48–49.
Acemogluo, Daron and Simon Johnson. *Power and Progress: Our Thousand-Year Struggle over Technology and Prosperity.* London: Basic Books, 2023.
Adami, Christoph. *The Evolution of Biological Information: How Evolution Creates Complexity, from Viruses to Brains.* Princeton NJ: Princeton University Press. 2024.
Agamben, Giorgio. *The Open: Man and Animal,* trans. Kevin Attell. Stanford CA: Stanford University Press, 2004.
Agrawal, Ajay, Joshua Gans, and Avi Goldfarb. *Prediction Machines: The Simple Economics of Artificial Intelligence.* Boston: Harvard Business Press, 2018.
Anderson, Chris, "The End of Theory: The Data Deluge Makes the Scientific Method Obsolete," *Wired* 16:1 (2008) https://www.wired.com/2008/06/pb-theory/.
Ansell Pearson, Keith. "Viroid Life: On Machines, Technics, and Evolution" in Keith Ansell Pearson (ed.), *Deleuze and Philosophy: The Difference Engineer,* 180–210. New York: Routledge 1997.
Araujo, Luis, and Debbie Harrison. "Path dependence, agency and technological evolution." *Technology Analysis & Strategic Management* 14:1 (2002): 5–19.
Asimov, Isaac. *The Complete Stories.* Crown, 1990.
Asimov, Isaac. "Foreword" in Pierre de Latil, *Thinking by Machine: A Study of Cybernetics,* trans. Y.M. Golla, v–viii. Boston MA: Houghton Mifflin, 1957.
Bakker, R. Scott. "Crash space." *Midwest Studies in Philosophy* 39 (2015), 186–204.

Barbrook, R. and Cameron, A. "The Californian Ideology". *Science as culture*, 6:1 (1996): 44–72.

Bardin, Andrea and Marco Ferrari. "Governing Progress: From Cybernetic Homeostasis to Simondon's Politics of Metastability". *The Sociological Review Monographs* 70:2 (2022): 248–263.

Basalla, George. *The Evolution of Technology*. New York: Cambridge University Press, 1988.

Beaumont, Thomas. "Musk ascends as a political force beyond his wealth by tanking budget deal." *Associated Press*. Dec. 19 2024. https://apnews.com/article/elon-musk-congress-bipartisan-deal-320a3487d596ae0d2e1ffdc1e620 54b3. Accessed Dec. 20 2024.

Beck, Naomi. "Be Fruitful and Multiply: Growth, Reason, and Cultural Group Selection in Hayek and Darwin." *Biological Theory* 6 (2011): 413–423.

Beckett, Andy "Accelerationism: how a fringe philosophy predicted the world we live in" *The Guardian*, May 11 2017 (https://www.theguardian.com/world/2017/may/11/accelerationism-how-a-fringe-philosophy-predicted-the-future-we-live-in).

Beer, David. *The Tensions of Algorithmic Thinking: Automation, Intelligence and the Politics of Knowing*. Bristol UK: University of Bristol Press, 2022.

Beer, Stafford. *The Brain of the Firm: The Managerial Cybernetics of Organization*. 2nd ed. New York: John Wiley & Sons, 1982.

Beer, Stafford. "What is Cybernetics?" *Kybernetes* 31:2 (2002): 209–219.

Bernstein, J.M. *Adorno: Disenchantment and Ethics*. New York: Cambridge University Press 2001.

Boden, Margaret A. *Mind as Machine: A History of Cognitive Science*. 2 vols. New York: Oxford University Press, 2006.

Boettke, Peter J., and Rosolino A. Candela. "On the feasibility of technosocialism." *Journal of Economic Behavior & Organization* 205 (2023): 44–54.

Boltanski, Luc and Ève Chiapello. *The New Spirit of Capitalism*, trans. Gregory Elliott. New York: Verso 2007.

Brandom, Robert B. "An Arc of Thought: From Rorty's Eliminative Materialism to His Pragmatism" in *Richard Rorty: From Pragmatist Philosophy to Cultural Politics*, ed. Alexander Gröschner, Colin Koopman, and Mike Sandbothe, 23–30. New York: Bloomsbury, 2013.

Brandom, Robert B. *Articulating Reasons: An Introduction to Inferentialism*. Cambridge MA: Harvard University Press. 2000.

Brassier, Ray. "Concrete-in-Thought, Concrete-in-Act: Marx, Materialism, and the Exchange Abstraction." *Crisis & Critique* 5:1 (2018a), 111–129.

Brassier, Ray. "Correlation, Speculation, and the Modal Kant-Sellars Thesis" in Fabio Gironi (ed.), *The Legacy of Kant in Sellars and Meillassoux: Analytic and Continental Kantianism*, 67–84. New York: Routledge, 2018b.

Brassier, Ray. *Nihil Unbound: Enlightenment and Extinction*. Palgrave Macmillan 2007.
Brassier, Ray. "Naturalism, Nominalism, and Materialism: Sellars' Critical Ontology" in Bana Bashour and Hans D. Muller (ed.), *Contemporary Philosophical Naturalism and Its Implications*, 101–114. New York: Routledge 2014a.
Brassier, Ray. "Prometheanism and Its Critics" in #*Accelerate: An Accelerationist Reader*, ed. Robin Mckay and Armen Avanessian, 467–487. Urbanomic 2014b.
Brassier, Ray. "Strange Sameness: Hegel, Marx, and the Logic of Estrangement." *Angelaki* 24:1 (2019), 98–105.
Brassier, Ray. "The View from Nowhere". *Identities: Journal for Politics, Gender and Culture* 8:2 (2011), 7–23.
Brown, Gordon S. and Nobert Wiener, "Automation, 1955: A Retrospective". *Annals of the History of Computing* 6:4 (1985), 372–383.
Burrell, Jenna. "How the Machine 'Thinks': Understanding Opacity in Machine Learning Algorithms." *Big Data & Society* 3:1 (2016), 1–12.
Cabrera, F. "The Fate of Explanatory Reasoning in the Age of Big Data." *Philosophy and Technology* 34 (2021): 645–665.
Caldwell, Bruce and Leonidas Montes. "Friedrich Hayek and his visits to Chile". *Review of Austrian Economics* 28 (2015).
Callaway, Ewen. "'It will change everything': DeepMind's AI makes gigantic leap in solving protein structures." *Nature* 588:7837 (2020), 203–205.
Cardon, Dominique, Jean-Philippe Cointet et Antoine Mazières, "La revanche des neurones. L'invention des machines inductives et la controverse de l'intelligence artificielle" *Réseaux* 5:11 (2018): 173–220.
Carroll, Lewis. "What the Tortoise said to Achilles." *Mind* 4:14 (1895), 278–280.
Caygill, Howard. "The Topology of Selection: The Limits of Deleuze's Biophilosophy" in Keith Ansell Pearson (ed.), *Deleuze and Philosophy: The Difference Engineer*, 149–162. New York: Routledge, 1997.
Celis Bueno, Claudio. "The face revisited: Using Deleuze and Guattari to explore the politics of algorithmic face recognition." *Theory, Culture & Society* 37:1 (2020): 73–91.
Chalmers, David. "The Singularity: A Philosophical Analysis". *Journal of Consciousness Studies* 7:9–10 (2010): 7–65.
Chen, Boyuan, Kuang Huang, Sunand Raghupathi, Ishaan Chandratreya, Qiang Du, and Hod Lipson. "Automated discovery of fundamental variables hidden in experimental data." *Nature Computational Science* 2:7 (2022), 433–442.
Christias, Dionysis. *Normativity, Lifeworld and Science in Sellars' Synoptic Vision*. New York: Palgrave Macmillan. 2023.
Churchland, Paul. "On the Nature of Theories: A Neurocomputational Perspective" in *Mind Design II: Philosophy, Psychology, Artificial Intelligence*,

Revised and expanded edition, ed. John Haugeland, 251–292. Cambridge MA: The MIT Press 1997.

Clark, Andy. *Natural Born Cyborgs: Minds, Technologies, and the Future of Human Intelligence*. New York: Oxford University Press, 2003.

Coeckelbergh, Mark. *Using Words and Things: Language and Philosophy of Technology*. New York: Routledge. 2017.

Connolly, William. *The Fragility of Things: Self-Organizing Processes, Neoliberal Fantasies, and Democratic Activism*. Durham NC: Duke University Press 2013.

Cooper, Melinda. *Family Values: Between Neoliberalism and the New Social Conservatism*. New York: Zone Books, 2017.

Clynes, Manfred and Nathan Kline. "Cyborgs and Space," *Astronautics* 14:9 (1960), 26–27.

Dafoe, A. "On technological determinism: A typology, scope conditions, and a mechanism." *Science, Technology, & Human Values*, 40:6 (2015), 1047–1076.

Dapprich, Jan Philipp and Dan Greenwood. "Cybersocialism and the Future of the Socialist Calculation Debate." *Erasmus Journal for Philosophy and Economics* 17:1 (2024): 1–23.

Dennett, Daniel. "True Believers: The Intentional Strategy and Why It Works" [1981] in John Haugeland (ed.), *Mind Design II: Philosophy, Psychology, Artificial Intelligence*, 57–81. Cambridge MA: The MIT Press 1997.

Deleuze, Gilles and Félix Guattari. *A Thousand Plateaus: Capitalism and Schizophrenia Vol. 2*, trans. Brian Massumi. Minneapolis MN: University of Minnesota Press, 1987.

Devezas, T. C. "Evolutionary theory of technological change: State-of-the-art and new approaches." *Technological forecasting and social change*, 729: (2005), 1137–1152.

D'Oro, Giuseppina. "How To (and How not To) Defend the Manifest Image" in Paul Giladi (ed.), *Responses to Naturalism: Critical Perspectives from Idealism and Pragmatism*, 144–164. New York: Routledge, 2019.

Dor, Daniel. *The Instruction of Imagination: Language as a Social Communicative Technology*. New York: Oxford University Press. 2015.

Dupuy, Jean-Pierre. *The mechanization of the mind: On the origins of cognitive science*. Princeton NJ: Princeton University Press, 2001.

Dryzek, John S. *The politics of the earth: Environmental discourses*. New York: Oxford University Press, 2013.

Dyer-Witheford, Nick, Atle Mikkola Kjøsen, and James Steinhoff. *Inhuman Power: Artificial intelligence and the Future of Capitalism*. London: Pluto Press, 2019.

Eklund, Matti. *Alien Structure: Language and Reality*. New York: Oxford, 2024.

Elliott, Anthony. *Psychoanalytic Theory: An Introduction*. Oxford: Blackwell 1994.

Ellul, Jacques. *The Technological Society*, trans. John Wilkinson. New York: Vintage Books. 1964.

Fisher, Mark. *Ghosts of my life: Writings on Depression, Hauntology and Lost Futures.* Winchester UK: Zero Books, 2014a.
Fisher, Mark. "Terminator vs Avatar" in *#Accelerate: An Accelerationist Reader*, ed. Robin Mckay and Armen Avanessian, 335–346. Falmouth UK: Urbanomic 2014b.
Floridi, Luciano. "Is Semantic Information Meaningful Data?" *Philosophy and Phenomenological Research* 70:2 (2005): 351–370.
Floridi, Luciano. "Semantic information and the correctness theory of truth." *Erkenntnis* 74 (2011): 147–175.
Foley, Richard. "Dretske's 'information-theoretic' account of knowledge." *Synthese* 70:2 (1987): 159–184.
Foucault, Michel. *The Birth of Biopolitics: Lectures at the Collège de France 1978–1979*, trans. Graham Burchell, ed. Michel Senellart. New York: Palgrave Macmillan, 2008.
Foucault, Michel. *The Order of Things: An Archaeology of the Human Sciences.* New York: Routledge, 2002.
Foucault, Michel. *Security, Territory, Population: Lectures at the Collège de France 1977–1978*, trans. Graham Burchell, ed. Michel Senellart. New York: Palgrave Macmillan, 2007.
Foucault, Michel. "Technologies of the Self" in *Technologies of the Self: A Seminar with Michel Foucault*, ed. Luther H. Martin, Huck Gutman, Patrick H. Hutton, 16–49. Amherst MA: University of Massachusetts Press, 1989.
Fourcade, Marion, and Kieran Healy. "Seeing Like a Market." *Socio-economic Review* 15:1 (2017), 9–29.
Galison, Peter. "The Ontology of the Enemy: Norbert Wiener and the Cybernetic Vision." *Critical Inquiry* 21:1 (1994): 228–266.
Gamez, Patrick. "Being Truly Wrong: Enlightened Nihilism or Unbound Naturalism?" *Open Philosophy* 6 (2023a), 1–28.
Gamez, Patrick. "A Friendly Critique of Levinasian Machine Ethics". *Southern Journal of Philosophy* 60:1 (2022), 118–149.
Gamez, Patrick. "Inhuman Hermeneutics of the Self: Biopolitics in the Age of Big Data." *Foucault Studies* 34 (2023b): 80–110.
Gamez, Patrick. "Metaphysics or Metaphors for the Anthropocene: Scientific Naturalism and the Agency of Things." *Open Philosophy* 1 (2018): 191–212.
Gamez, Patrick. "The Place of the Iranian Revolution in the History of Truth: Foucault on Neoliberalism, Spirituality, and Enlightenment." *Philosophy and Social Criticism* 45:1 (2019): 2019, 96–124.
Gamez, Patrick. "On Being Born Poorly: Steps Toward a Genuinely Postvital Posthumanism." *Philosophy Today*, forthcoming.
Gardiner, Michael E. "Critique of accelerationism." *Theory, Culture & Society* 34:1 (2017), 29–52.

Gaus, Gerald F. *On Philosophy, Politics, and Economics*. Belmont CA: Wadsworth Cengage Learning, 2008.
Gebru, Timnit, and Émile P. Torres. "The TESCREAL Bundle: Eugenics and the Promise of Utopia through Artificial General Intelligence". *First Monday* 29:4 (2024). https://doi.org/10.5210/fm.v29i4.13636.
Godfrey-Smith, Peter. 2007. "Information in biology." In *The Cambridge Guide to the Philosophy of Biology*, ed. David L. Hull and Michael Ruse, 103–119. Cambridge: Cambridge University Press.
Golumbia, David. "Hypercapital." *Postmodern Culture* 7:1 (1996).
Günther, Mario, and Atoosa Kasirzadeh. "Algorithmic and human decision making: for a double standard of transparency." *AI & SOCIETY* 37 (2022), 375–381.
Gutting, Gary. *Thinking the Impossible: French Philosophy since 1960*. New York: Oxford University Press, 2013.
Halpern, Orit, and Robert Mitchell. *The Smartness Mandate*. Cambridge MA: The MIT Press, 2023.
Hancock, David. *The Countercultural Logic of Neoliberalism*. New York: Routledge, 2019.
Hancock, Max. "Spontaneity and Control: Friedrich Hayek, Stafford Beer, and the Principles of Self-Organization" *Modern Intellectual History*. Published online 2024. https://doi.org/10.1017/S1479244324000076
Hausman, Daniel M. "Lessons from Quantum Mechanics." *Synthese* 121 (1999): 79–92.
Hayek, Friedrich. "Competition as a Discovery Procedure" in *The Market and Other Orders: The Collected Works of F.A. Hayek Volume XV*, edited by Bruce Caldwell, 304–313. Chicago: University of Chicago Press, 2014a.
Hayek, Friedrich. "The Primacy of the Abstract" in *The Market and Other Orders: The Collected Works of F.A. Hayek Volume XV*, edited by Bruce Caldwell, 314–327. Chicago: University of Chicago Press, 2014b.
Hayek, Friedrich. *The Road to Serfdom*. New York: Routledge Classics, 2001.
Heidegger, Martin. "The end of philosophy and the task of thinking." In *Basic Writings*, ed. David Farrell Krell, 427–449. San Francisco: HarperSanFrancisco, 1993.
Heidegger, Martin. *Four Seminars*. Bloomington IN: Indiana University Press, 2012.
Hermansson, Patrik, David Lawrence, Joe Mulhall, and Simon Murdoch. *The International Alt-Right: Fascism for the 21st Century?* New York, Routledge, 2020.
Herrit, Robert. "Google's Philosopher". *Pacific Standard*, Dec. 30 2014 (Updated June 14 2017). https://psmag.com/environment/googles-philosopher-technology-nature-identity-court-legal-policy-95456 (Accessed June 2024).
Hester, Helen. "Sapience+ care: reason and responsibility in posthuman politics". *Angelaki* 24:1 (2019), 67–80.

Hicks, Michael Townsen, James Humphries, and Joe Slater. "ChatGPT is bullshit." *Ethics and Information Technology* 26 (2024). https://doi.org/10.1007/s10676-024-09775-5.
Hughes, T. "Technological Momentum" in *Technology and Society: Building Our Sociotechnical Future*, ed. Deborah G. Johnson and Jameson M. Wetmore, 141–150. Cambridge MA: The MIT Press 2009.
Hyppolite, Jean. *Logic and Existence*, trans. Leonard Lawlor and Amit Sen. Albany NY: SUNY Press 1997.
Jacob, François. *The Logic of Life: A History of Heredity*, trans. Betty E. Spillmann. Princeton NJ: Princeton University Press, 2022.
Jasanoff, Sheila. "Future Imperfect: Science, Technology, and the Imagination of Modernity" in Sheila Jasanoff & Sang Hyun-Kim (eds.), *Dreamscapes of Modernity: Sociotechnical Imaginaries and the Fabrication of Power*, 1–33. Chicago: University of Chicago Press, 2015.
Kasenberg, Daniel, Thomas Arnold, and Matthias Scheutz. "Norms, rewards, and the intentional stance: Comparing machine learning approaches to ethical training" In *Proceedings of the 2018 AAAI/ACM Conference on AI, Ethics, and Society* (2018), 184–190.
Kay, Lily E. *Who Wrote the Book of Life? A History of the Genetic Code*. Stanford CA: Stanford University Press, 2000.
Keller, Evelyn Fox. *Making Sense of Life: Explaining Biological Development with Models, Metaphors, and Machines*. Cambridge MA: Harvard University Press, 2003.
Kline, Ronald R. *The Cybernetics Moment, or Why We Call Our Age the Information Age*. Baltimore MD: Johns Hopkins University Press, 2015.
Koster, Jan. "Ceaseless, unpredictable creativity: Language as technology". *Biolinguistics* 3:1 (2009): 61–92.
Lambert, Karras J., and Tate Fegley. "Economic calculation in light of advances in big data and artificial intelligence." *Journal of Economic Behavior & Organization* 206 (2023): 243–250.
Lance, Mark Norris. "Placing in a Space of Norms: Neo-Sellarsian Philosophy in the 21st Century" in *The Oxford Handbook of American Philosophy*, ed. Cheryl Misak, 403–239. New York: Oxford University Press, 2008.
Land, Nick. "Cybergothic" in *Fanged noumena: Collected writings 1987–2007*, ed. Robin Mackay and Ray Brassier, 345–374. Falmouth UK: Urbanomic, 2011a.
Land, Nick. "The Dark Enlightenment" (https://www.thedarkenlightenment.com/the-dark-enlightenment-by-nick-land/).
Land, Nick. "Machines and Technocultural Complexity: The Challenge of the Deleuze-Guattari Conjunction." *Theory, Culture and Society* 12 (1995): 131–140.
Land, Nick. "Machinic Desire" in *Fanged noumena: Collected writings 1987–2007*, ed. Robin Mackay and Ray Brassier, 319–344. Falmouth UK: Urbanomic, 2011b.

Land, Nick. "Making It With Death: Remarks on Thanatos and Desiring-Production" in *Fanged noumena: Collected writings 1987–2007*, ed. Robin Mackay and Ray Brassier, 261–288. Falmouth UK: Urbanomic, 2011c.
Land, Nick. "Meltdown" in *Fanged noumena: Collected writings 1987–2007*, ed. Robin Mackay and Ray Brassier, 441–460. Falmouth UK: Urbanomic, 2011d.
Land, Nick. "Teleoplexy: Notes on Acceleration" in *#Accelerate: An Accelerationist Reader*, ed. Robin Mckay and Armen Avanessian, 509–520. Urbanomic 2014.
Lange, Oskar. "The computer and the market" in *Socialism, capitalism and economic growth: Essays presented to Maurice Dobb*, pp. 158–161. Cambridge: Cambridge University Press, 1967.
Lombardi, Olimpia, Federico Holik, and Leonardo Vanni. "What is Shannon information?" *Synthese* 193 (2016): 1983–2012.
Lyotard, Jean-Francois. *Libidinal Economy*, trans. Iain Hamilton Grant. Bloomington IN: Indiana University Press 1993.
Mackay, Robin. "Nick Land: An Experiment in Inhumanism." http://readthis.wtf/writing/nick-land-an-experiment-in-inhumanism/.
Malafouris, Lambros. *How Things Shape the Mind: A Theory of Material Engagement*. Cambridge MA: The MIT Press, 2013.
March, James G. and Herbert A. Simon. *Organizations*. New York : John Wiley and Sons, 1958.
Marchesi, S., Ghiglino, D., Ciardo, F., Perez-Osorio, J., Baykara, E. and Wykowska, A.,. "Do we adopt the intentional stance toward humanoid robots?" *Frontiers in Psychology* 10 (2019): https://doi.org/10.3389/fpsyg.2019.00450.
Mayr, Otto. *Authority, Liberty, and Automatic Machinery in Early Modern Europe*. Cambridge MA: The MIT Press, 1986.
Mayr, Otto. *The Origins of Feedback Control*. Cambridge MA: The MIT Press, 1970.
Mazzochi, F. "Could Big Data Be the End of Theory in Science." *EMBO Rep 16* (2015), 1250–1255.
Medina, Eden. *Cybernetic Revolutionaries: Technology and Politics in Allende's Chile*. Cambridge MA: The MIT Press, 2011.
McDowell, John. *Having the World in View: Essays on Kant, Hegel, and Sellars*. Cambridge MA: Harvard University Press, 2009.
McDowell, John. *Mind and World*. Cambridge MA: Harvard University Press. 1996.
McQuillan, Dan. *Resisting AI: An Anti-fascist Approach to Artificial Intelligence*. Bristol UK: University of Bristol Press, 2022.
Millière, Raphaël. "Philosophy of cognitive science in the age of deep learning." *Wiley Interdisciplinary Reviews: Cognitive Science* 15:4 (2024): e1684.
Millière, Raphaël, and Cameron Buckner. "A Philosophical Introduction to Language Models--Part I: Continuity With Classic Debates." *arXiv preprint arXiv:2401.03910* (2024).

Mirowski, Philip. "The Future (s) of Open Science." *Social Studies of Science* 48:2 (2018), 171–203.

Mirowski, Philip. "Postface: Defining Neoliberalism" in *The Road from Mont Pèlerin: The Making of the Neoliberal Thought Collective*, ed. Philip Mirowski and Dieter Plehwe, 417–456. Cambridge MA: Harvard University Press, 2009.

Moldbug, Mencius. "AGW, KFM, and HNU" in *Unqualified Reservations Volume One: A Gentle Introduction to Reactionary Enlightenment*, 33–62. New York: TRO LLC, 2009.

Mouffe, Chantal. "Carl Schmitt and the Paradox of Liberal Democracy." *Canadian Journal of Law & Jurisprudence* 10:1 (1997), 21–33.

Mouffe, Chantal. "Radical democracy or liberal democracy?" in David Trend (ed.), *Radical democracy: Identity, Citizenship and the State*, 19–26. New York: Routledge. 2013.

Mufwene, Salikoko S. "Language as technology: Some questions that evolutionary linguistics should address," in *The Search for Universal Grammar: From Old Norse to Zoque*, ed. Terje Lohndal, 327–358. Philadelphia PA: John Benjamins Publishing. 2013.

Müller, Vincent C., and Michael Cannon. "Existential risk from AI and orthogonality: Can we have it both ways?." *Ratio* 35:1 (2022), 25–36.

Negarestani, Reza. *Intelligence and Spirit*. Falmouth UK: Urbanomic, 2018.

Negarestani, Reza. "The Labor of the Inhuman" in *#Accelerate: An Accelerationist Reader*, ed. Robin Mckay and Armen Avanessian, 425–466. Urbanomic 2014.

Negarestani, Reza. "Revolution Backwards: Functional Realization and Computational Implementation" in *Alleys of Your Mind: Augmented Intelligence and Its Traumas*, ed. Matteo Pasquinelli. Milton Keynes, UK: meson press, 2015.

Nietzsche, Friedrich. *On the Genealogy of Morality*, trans. Carol Diethe. Ed. Keith Ansell-Pearson. Cambridge University Press 2006.

Nik-Khah, Edward, and Philip Mirowski. "The ghosts of Hayek in orthodox microeconomics: Markets as information processors" in *Markets*, 31–70. Minneapolis MN: University of Minnesota Press 2019.

Norris, Benjamin. "The Education of Kanzi and the Notion of Progress: Reza Negarestani's *Intelligence and Spirit*." In *Pli* 30 (2019), 164–190.

Norton, John. "Causation as Folk Science," in *Causation, Physics, and the Constitution of Reality: Russell's Republic Revisited*, ed. Huw Price and Richard Corry, 11–44. New York: Oxford University Press, 2007.

Noys, Benjamin. *Malign Velocities: Accelerationism and Capitalism*. Winchester UK: Zero Books 2015.

Noys, Benjamin. *Persistence of the Negative: A Critique of Contemporary Continental Theory*. Edinburgh UK: Edinburgh University Press, 2010.

Okrent, Mark. "On Layer Cakes: Heidegger's Normative Pragmatism Revisited" in Ondřej Švec and Jakub Čapek (eds.), *Pragmatic Perspectives in Phenomenology*, 21–38. New York: Routledge, 2017.

Oliva, Gabriel. "The Road to Servomechanisms: The Influence of Cybernetics on Hayek from The Sensory Order to the Social Order". CHOPE Working Paper, No. 2015-11. Durham, NC: Duke University, Center for the History of Political Economy (CHOPE), 2015.

Olsson, Carl. "Peak Face". *Urbanomic Documents* UFD0056. https://www.urbanomic.com/wp-content/uploads/2023/06/Urbanomic_Documents_UFD0056_Peak_face.pdf. Accessed Dec. 30 2024.

Pasquinelli, Matteo. "How to Make a Class: Hayek's Neoliberalism and the Origins of Connectionism." *Qui Parle* 30:1 (2021): 159–184.

Paul, L.A. *Transformative Experience*. New York: Oxford University Press, 2014.

Perez-Osorio, Jairo, and Agnieszka Wykowska. "Adopting the intentional stance toward natural and artificial agents." *Philosophical Psychology* 33:3 (2020): 369–395.

Pietsch, Wolfgang. *On the Epistemology of Data Science: Conceptual Tools for a New Inductivism*. Cham, CH: Springer, 2021.

Pigliucci, Massimo. "The end of theory in science?." *EMBO reports* 10:6 (2009), 534–534.

Pogue, James. "Inside the New Right, Where Peter Thiel is Placing His Biggest Bets." *Vanity Fair*, April 20 2022. https://www.vanityfair.com/news/2022/04/inside-the-new-right-where-peter-thiel-is-placing-his-biggest-bets. Accessed Dec. 20 2024.

Quine, W.V. "Epistemology Naturalized" in *Ontological Relativity and Other Essays*. New York: Columbia University Press, 1969.

Ramsey, William, Stephen Stich, and Joseph Garon, "Connectionism, Eliminativism, and the Future of Folk Psychology" in *Mind Design II: Philosophy, Psychology, Artificial Intelligence, Revised and expanded edition*, ed. John Haugeland, 351–376. Cambridge MA: The MIT Press 1997.

Roden, David. "The Filter Problem for Bioethics: The Case of Hyperagency" in Danielle Sands (ed.), *Bioethics and the Posthumanities*, 116–128. New York: Routledge, 2022.

Roden, David. *Posthuman Life: Philosophy at the Edge of the Human*. New York: Routledge. 2015.

Roden, David. "On Reason and Spectral Machines: Robert Brandom and Bounded Posthumanism" in Rosi Braidotti and Rick Dolphijn (eds.), *Philosophy after Nature*, 99–120. New York: Rowman & Littlefield. 2017.

Roden, David. "Subtractive-Catastrophic Xenophilia." *Identities: Journal for Politics, Gender and Culture* 16:1–2 (2019), 40–46.

Rorty, Richard. "In Defense of Eliminativism." *The Review of Metaphysics* 24:1 (1970): 112–121.

Rothbard, Murray. "A Strategy for the Right." Rothbard-Rockwell Report 3:3 (1992a): 1–16.
Rothbard, Murray. "Right-Wing Populism: A Strategy for the Paleo Movement." Rothbard-Rockwell Report 3:1 (1992b): 5–14.
Rouvroy, Antoinette and Bernard Stiegler. "The Digital Regime of Truth: From the Algorithmic Governmentality to a New Rule of Law". La Deleuziana 3 (2016): 6–27.
Rycroft, Robert W., and Don E. Kash. "Path dependence in the innovation of complex technologies." Technology Analysis & Strategic Management 14:1 (2002): 21–35.
Sachs, Carl B. "Normativity, Lifeworld, and Science in Sellars' Synoptic Vision". International Journal of Philosophical Studies 31:5 (2023): 739–744.
Sandifer, Elizabeth. Neoreaction a Basilisk. Eruditorum Press, 2017.
Sejnowski, Terrence J. The Deep Learning Revolution: Machine Intelligence meets Human Intelligence. Cambridge MA: The MIT Press. 2018
Sejnowski, Terrence J. ChatGPT and the Future of AI: The Deep Language Revolution. Cambridge MA: The MIT Press. 2024.
Sellars, Wilfrid. "Being and Being Known," in Science, Perception, and Reality, 41–59. Atascadero, CA: Ridgeview. 1963a.
Sellars, Wilfrid. Empiricism and the Philosophy of Mind, ed. Richard Rorty, with a Study Guide by Robert Brandom. Cambridge MA: Harvard University Press. 1997.
Sellars, Wilfrid. "Language, Rules, and Behavior" in Sidney Hook (ed.), John Dewey: Philosopher of Science and Freedom, 289–315. New York: The Dial Press. 1949.
Sellars, Wilfrid. Naturalism and Ontology: The John Dewey Lectures of 1974. Atascadero CA: Ridgeview. 1979.
Sellars, Wilfrid. "Philosophy and the Scientific Image of Man" in Science, Perception, and Reality, 1–40. Atascaedro CA: Ridgeview. 1963b.
Shapiro, Lionel. "Sellars, Truth Pluralism, and Truth Relativism" in Stefan Brandt and Anke Breunig (ed.), Wilfrid Sellars and Twentieth-Century Philosophy, 174–206. New York: Routledge, 2020.
Shear, Michael D. et al. "Elon Musk Flexes His Political Muscle as Government Shutdown Looms." The New York Times. Dec. 19 2024. https://www.nytimes.com/2024/12/19/us/politics/elon-musk-politics.html. Accessed Dec. 20 2024.
Simon, Herbert A. The Sciences of the Artificial, 3rd ed. Cambridge MA: The MIT Press, 2019.
Simondon, Gilbert. On the Mode of Existence of Technical Objects, trans. Cecile Malaspina and John Rogove. Minneapolis MN: Univocal, 2017.

Slobodian, Quinn. "Anti-'68ers and the Racist-Libertarian alliance: How a Schism among Austrian School Neoliberals Helped Spawn the Alt Right." *Public Culture*, 15:3 (2019): 372–386.

Slobodian, Quinn. *Crack-Up Capitalism: Market Radicals and the Dream of a World without Democracy*. New York: Metropolitan Books 2023a.

Slobodian, Quinn. "The Ethno-economy: Peter Brimelow and the Capitalism of the Far Right." *Journal of American Studies* (2024). https://doi.org/10.1017/S002187582400015X.

Slobodian, Quinn. "The Unequal Mind: How Charles Murray and Neoliberal Think Tanks Revived IQ." *Capitalism: A Journal of History and Economics* 4:1 (2023b), 73–108.

Sorabji, Richard. *Emotion and Peace of Mind from Stoic Agitation to Christian Temptation*. New York: Oxford University Press, 2000.

Smith, Gary, and Jay Cordes. *The phantom pattern problem: The mirage of big data*. Oxford University Press, 2020.

Srnicek, Nick. *Platform Capitalism*. Malden MA: Polity 2017.

Srnicek, Nick, and Alex Williams. *Inventing the Future: Postcapitalism and a World without Work*. New York: Verso Books, 2015.

Steinhoff, *Automation and Autonomy: Labour, Capital and Machines in the Artificial Intelligence Industry*. Cham, Switzerland: Springer 2021.

Stich, Stephen P. "From connectionism to eliminativism." *Behavioral and Brain Sciences* 11:1 (1988): 53–54.

Stiegler, Bernard. *Automatic Society Vol. 1: The Future of Work*, trans. Daniel Ross. Malden MA: Polity, 2016.

Stiegler, Bernard. *Nanjing Lectures: 2016–2019*. Edited and translated by Daniel Ross. London: Open Humanities Press, 2020.

Stiegler, Bernard. *The Neganthropocene*. Edited and translated by Daniel Ross. London: Open Humanities Press, 2018.

Sunstein, Cass. "Hayekian Behavioral Economics." *Behavioural Public Policy* 7 (2023): 170–188.

Sunstein, Cass and Richard A. Thaler. *Nudge: Improving Decisions about Health, Wealth, and Happiness*. New Haven CT: Yale University Press, 2008.

Thaler, Richard A. *Misbehaving: The Making of Behavioural Economics*. London: Penguin, 2015.

Thellman, Sam, Maartje De Graaf, and Tom Ziemke. "Mental state attribution to robots: A systematic review of conceptions, methods, and findings." *ACM Transactions on Human-Robot Interaction* 11:4 (2022): 1–51.

Thompson, Michael J. *The Twilight of the Self: The Decline of the Individual in Late Capitalism*. Stanford CA: Stanford University Press, 2022.

Toosi, A. et al. "A brief history of AI: how to prevent another winter (a critical review)." *PET clinics* 16:4 (2021): 449–469.

Torres, Émile P. *Human Extinction: A History of the Science and Ethics of Annihilation*. New York: Routledge. 2023.
Turing, Alan. "Computing Machinery and Intelligence". *Mind* 59:236 (1950), 433–460.
Vallor, Shannon. *The AI Mirror: How to Reclaim Our Humanity in an Age of Machine Thinking*. New York: Oxford University Press, 2024.
van Fraassen, Bas. "The Charybdis of Realism: Epistemological Implications of Bell's Inequality." *Synthese* 52 (1982a): 25–38.
van Fraassen, Bas. "On the Radical Incompleteness of the Manifest Image." In *PSA: Proceedings of the Biennial Meeting of the Philosophy of Science Association*, 1976:2, 335–343. Cambridge University Press. 1976.
van Fraassen, Bas. "Rational Belief and the Common Cause Principle" in *What? Where? When? Why? Essays on Induction, Space and Time, Explanation*, ed. R. McLaughlin, 193–210. Dordrecht, NL: D. Reidel Publishing, 1982b.
van Fraassen, Bas. *The Scientific Image*. New York: Oxford University Press 1980.
van Fraassen, Bas. "Wilfrid Sellars' apocalyptic vision". https://basvanfraassensblog.home.blog/2023/08/13/wilfrid-sellars-apocalyptic-vision/. Accessed June 25 2024.
Voss, Daniela. "Gilbert Simondon and Difference Senses of 'Evolution'". *Angelaki* 29:5 (2024): 97–113.
Wallace-Wells, Benjamins. "The Rise of the Thielists." *The New Yorker*. May 13 2021. https://www.newyorker.com/news/annals-of-populism/the-rise-of-the-thielists. Accessed Dec. 20 2024.
Wolfendale, Peter. "The Reformatting of Homo Sapiens". *Angelaki: Journal of the Theoretical Humanities* 24:1 (2019), 55–66.
White, Joel. "How does one Cosmotheoretically Respond to the Heat Death of the Universe?" *Open Philosophy* 6:1 (2023): 20220233.
Whitehead, Aldred North and Bertrand Russell. *Principia Mathematic to *56*. New York: Cambridge University Press, 1997.
Wiener, Norbert. *Cybernetics: Control and Communication in the Animal and the Machine*. Cambridge MA: The MIT Press, 1961.
Williams, Alex. "Strategy without a Strategiser." *Angelaki* 24:1 (2019), 14–25.
Williams, Alex and Nick Srnicek. "#Accelerate: Manifesto for an Accelerationist Politics" in *#Accelerate: An Accelerationist Reader*, ed. Robin Mckay and Armen Avanessian, 347–462. Falmouth UK: Urbanomic 2014.
Williams, Damien Patrick. "Bias Optimizers." *American Scientist* 111:4 (2023): 204–207.
Winner, Langdon. *Autonomous Technology: Technics-out-of-Control as a Theme in Political Thought*. Cambridge MA: The MIT Press. 1977.
Wittgenstein, Ludwig. *Philosophical Investigations, Revised 4th Edition*, ed. P.M.S. Hacker and Joachim Schulte, trans. G.E.M. Anscombe, P.M.S. Hacker, and Joachim Schulte. Malden MA: Blackwell. 2009.

Zednik, Carlos. "Solving the black box problem: A normative framework for explainable artificial intelligence." *Philosophy & technology* 34:2 (2021), 265–288.

Zerilli, John. "Explaining machine learning decisions." *Philosophy of Science* 89:1 (2022), 1–19.

Zerilli, John, Alistair Knott, James Maclaurin, and Colin Gavaghan. "Transparency in algorithmic and human decision-making: is there a double standard?" *Philosophy & Technology* 32 (2019), 661–683.

Zuboff, Shoshana. *The Age of Surveillance Capitalism: The Fight for a Human Future at the New Frontier of Power.* New York: Public Affairs, 2019a.

Zuboff, Shoshana. "Big other: surveillance capitalism and the prospects of an information civilization." *Journal of Information Technology* 30:1 (2015), 75–89.

Zuboff, Shoshana. *In the Age of the Smart Machine: The Future of Work and Power.* New York: Basic Books, 1988.

Zuboff, Shoshana. "Surveillance Capitalism and the Challenge of Collective action." *New Labor Forum* 28:1 (2019b), 10–29.

Zuboff, Shoshana. "'We Make Them Dance': Surveillance Capitalism, the Rise of Instrumentarian Power, and the Threat to Human Rights." in Human rights in *the age of platforms*, ed. Rikke Frank Jorgensen, 3–51. Cambridge MA; The MIT Press 2019c.

Index[1]

A
Arendt, Hannah, 14, 14n6, 26
Artificial intelligence (AI), 2, 3, 7–9,
　21, 31, 32, 35, 40, 44, 46,
　74–76, 82–85, 88, 95–102,
　96n17, 97n19, 98n25, 103n37,
　105, 107, 108, 110–118,
　110n60, 118n82, 121–124,
　126–128, 132, 135, 136
　large language models, 97n19, 113
Asimov, Isaac, 2–4, 7–9, 13, 16, 19,
　27, 83, 102, 106, 132, 133
Automation, 84, 98n25, 99, 100,
　103, 110

B
Beer, Stafford, 105, 105n44
Behavioral economics, 112
　nudging, 110, 111

Belief, 51, 53, 56, 58, 58n30, 61, 70,
　73, 117, 122, 123, 125, 127,
　128, 130
Big Data, 96, 107, 108, 126,
　126n101, 133
　data imperative, 109, 119
Brandom, Robert, 18, 18n25, 45,
　54, 55, 69
Brassier, Ray, 12, 13, 14n6, 16,
　16n17, 19–21, 26, 27, 38, 45,
　51, 52n18, 57–60, 64, 65,
　72–74, 72n54, 78–80, 83, 91,
　132, 133, 135

C
Calculation debate, 103, 104, 108
Capitalism, 25, 29–31, 29n10, 33,
　39–41, 59, 79, 88, 89, 94, 95,
　98, 99, 101–113, 116, 117

---

[1] Note: Page numbers followed by 'n' refer to notes.

Capitalism (cont.)
  platform, 2, 21, 94, 101–113, 116, 117, 123, 124, 126, 129, 133, 135, 137
Christias, Dionysis, 12, 38, 38n36, 45, 47n6, 49, 52n19, 59, 60, 63, 65, 68, 72n54, 74, 78, 81, 91, 121
Connectionism, 98, 103, 122
Cybernetics, 2–7, 9, 14, 16, 19, 21, 24–33, 46, 53, 57, 66, 74–76, 78, 93n10, 94–102, 104–107, 105n44, 109–114, 121n89, 129, 132
  allostasis (heterostasis), 102
  autopoiesis, 6, 102
  feedback, 7, 9, 53, 67, 109, 129
  homeostasis, 102
  second-order, 102, 112
  systems theory, 102
Cybersyn, 104, 105, 108
Cyborg, 14, 38–42, 61, 78, 123

D
Data capitalism, see Capitalism, platform
Deleuze, Gilles, 24, 25, 27–29, 33, 34, 52n20, 92, 92n10, 133, 134
  faciality, 133
  and Félix Guattari, 24, 25, 27–29, 33, 52n20, 133, 134
  schizoanalysis, 24, 25
Democracy, 34, 41, 105
  voice, 34
Dennett, Daniel, 123, 126, 128
Desire, 1, 2, 24, 25, 27–35, 28n8, 29n10, 37, 38, 40, 42, 44, 45, 52, 52n20, 60, 61, 66, 70, 72, 73, 84, 90, 95, 100, 104, 106, 107, 112, 113, 115–117, 116n77, 122, 123, 125, 127, 129–132, 134, 136
  desiring-production, 29

Desiring-production, 27–33, 88
Digital capitalism, see Capitalism, platform
Dupuy, Jean-Pierre, 13–16, 14n6, 60

E
Eliminativism, 45, 51, 74, 117, 121, 122
Ellul, Jacques, 17, 90, 93
Eugenics, 35, 37
Evolution, 14, 31, 34, 67, 92–94, 102, 106, 110
Exit, 34, 35, 41–43, 136
Explanation, 19, 36, 47, 49, 50, 61, 66–71, 70n49, 91, 92, 122–124, 126–127n103, 128–130, 128n104, 129n106

F
Fisher, Mark, 23, 26, 27, 38–40, 88, 89, 91, 99, 103, 107
  hauntology, 39
Folk psychology, 28, 44, 70, 114
  humean, 28, 28n8, 117
Foucault, Michel, 52, 63, 78n67, 105, 106, 113, 114n72, 115, 116, 116n77, 125
Future, 2, 3, 7, 9, 11–13, 16–21, 24, 25, 27, 31, 32, 37–41, 42n45, 48, 57n28, 74, 76, 77, 82–85, 88, 90, 91, 93, 95, 96, 103, 108, 113, 126, 132, 135–137

G
Given, 14–16, 53, 54, 81
Governmentality, 105, 106, 112n69, 113, 125

INDEX 155

H
Hayek, Friedrich, 25, 30–32, 36, 97–106, 105n44, 108, 111, 112, 116
Heidegger, Martin, 14n6, 17, 21, 23–26

I
Intentionality, 18–19n25, 28, 53n21, 71, 100, 123
Intentional stance, 123–125, 128

K
Kapp, Ernst, 61

L
Land, Nick, 21, 21n32, 23–30, 26n6, 32–42, 38n36, 44, 45, 51, 52n20, 59, 72, 74, 83, 84, 90–92, 92n10, 95, 96n16, 97, 98, 101–105, 107, 113, 120, 134
accelerationism, 21, 32, 33, 38, 53
Lange, Oskar, 103–105, 108, 112
Latour, Bruno, 61
Lyotard, Jean-Francois, 29, 29n10, 95, 135
libidinal economy, 24, 29

M
Market, 27, 30–32, 35, 36, 88, 89, 97–109, 111, 112, 115–117
Marx, Karl, 27, 29, 39, 42n45, 79
Marxism, 24, 25, 29, 103
autonomist, 103
critique of political economy, 103
post-Marxism, 103–104
technological determinism, 88
McDowell, John, 53n21, 54, 69, 72

Mind, 1–9, 12, 17, 17n19, 19, 21, 24, 37, 44–47, 49–54, 52n20, 57n28, 61–63, 66, 67, 69, 73, 78, 81, 87–132, 137
extended, 61
functionalism, 45, 46, 61
Mirowski, Philip, 30, 31, 111
Moldbug, Mencius, *see* Yarvin, Curtis
Mont Pelerin Society, 76

N
Naturalism, 21, 46, 47, 50, 59
Negarestani, Reza, 12, 19, 26, 38, 45, 51, 52, 59, 74, 76, 81–84, 88, 94, 114, 120, 121, 127
Neoliberalism, 30, 32, 37, 41, 96–98, 100, 101, 105, 111–113, 115n74
Austrian, 37, 112
Neoreaction, 35
Dark Enlightenment, 35

O
Outside, 25, 38, 41–44, 82, 83, 88, 114, 136

P
Pasquinelli, Matteo, 31, 98, 98n25, 103
Personhood, 28, 49, 52, 52n18, 55, 58, 63, 67, 73, 134
Prediction, 9, 108–110, 112, 113, 117, 119, 123, 126–128, 132
Prometheanism, 2, 9, 12, 13, 13n3, 14n6, 19, 21, 31, 33, 38–40, 38n36, 42–85, 88, 90, 99, 106, 131, 137
Psychoanalysis, 24, 25, 96n17, 114, 115, 115n74

Q
Quine, W.V., 58, 117

R
Rationality, 2, 12, 13n3, 17–21, 18n23, 28, 31, 32, 41, 45, 47, 48, 52, 53, 56–59, 62, 63, 67, 68, 72, 76, 78, 78n67, 81–84, 97, 111, 113, 115, 117, 121, 123–125, 127, 132
Roden, David, 45, 57n28, 89, 90, 93, 135, 136n9
Rorty, Richard, 69, 122
Rothbard, Murray, 36, 116

S
Schrödinger, Erwin, 7
negentropy, 7
Self-organization, 96, 99, 100n33, 111, 117
Sellars, Wilfrid, 21, 45–55, 50n13, 52n18, 53n21, 58, 59, 61, 63–65, 67–72, 70n49, 72n54, 72n55, 74–76, 78, 121, 122, 126, 126n100
manifest image, 46–52, 50n13, 54, 58, 63, 68–72, 70n49, 74, 75, 122, 126n100
myth of the given, 46, 53–55, 69
picturing, 71, 72, 72n55, 74, 75
Shannon, Claude, 4–6
Space of reasons, 46, 51–64, 66, 68–72, 77, 78, 95, 123, 127
game of giving and asking for reasons, 18, 45n2, 52, 53, 64, 65, 69, 72, 76, 78
Srnicek, Nick, 12, 38, 40, 104, 107n51
Stiegler, Bernard, 14–17, 15n10, 20, 21, 61, 92n10, 132
Surveillance capitalism, *see* Capitalism, platform

T
Technology, 2, 11–15, 17–19, 18n24, 24, 25, 36, 40, 40n41, 42, 44, 61, 85, 87–95, 93n12, 96n17, 98, 98n22, 99, 101, 103, 104, 106, 107, 109, 113, 114, 114n72, 119, 127, 132, 134
technological determinism, 77, 88, 90, 91
Temporality, 74, 80, 83, 84, 135
Thermodynamics, 2–4, 3n2, 6, 8, 16, 32, 34, 91, 102, 132
entropy, 2–9, 21, 31, 32, 91, 105, 106, 113, 129, 132, 133, 135
Transhumanism, 2, 37, 38, 61, 88
Truth-telling, 114, 117, 130
self-knowledge, 116–117
veridiction, 117

V
van Fraassen, Bas, 47, 50n13, 126n100, 129
von Mises, Ludwig, 36, 103, 116

W
Williams, Alex, 12, 38, 40, 76, 104, 104n42
Wolfendale, Peter, 12, 19, 26, 38, 45, 52, 57, 62, 63, 65, 74, 77, 78n67, 94, 121

Y
Yarvin, Curtis, 32, 37, 38

Z
Zuboff, Shoshanna, 107, 107n51, 109, 112, 114, 119

The manufacturer's authorised representative in the EU is Springer Nature Customer Service Centre GmbH, Europaplatz 3, 69115 Heidelberg, Germany. If you have any concerns regarding our products, please contact ProductSafety@springernature.com

Printed and bound by CPI Group (UK) Ltd, Croydon, CR0 4YY
26/03/2026
02078951-0006